A Military Mustang

The Extraordinary Life of Captain John W. Arens

A Military Mustang

The Extraordinary Life of Captain John W. Arens

John W. Arens
and
Charley Valera

 **Anchor Book Press · Palatine**

A Military Mustang:
The Extraordinary Life of Captain John W. Arens
Copyright © 2019 John W. Arens & Charley Valera
Imprint: Anchor Book Press
440 W Colfax Street, Unit 1132, Palatine, IL 60078
ISBN: 9781949109276
Printed in the United States

All rights reserved. No part of this publication may be reproduced, stored in a retrieval system, or transmitted in any form by any means – electronic, mechanical, photocopy, recording, or any other – except for brief quotations, without the prior written permission of the author.

ALSO BY CHARLEY VALERA
My Father's War:
Memories from Our Honored WWII Soldiers

PRAISE FOR CHARLEY

"Woven together, albeit individually described, they were totally engaging as a linked story, from beginning to end."
    -Mrs. Joanne Holbrook Patton, Widow of Major General George S. Patton IV

"*My Father's War* is a must read not just for World War II veterans and history buffs, but for anyone who is moved by a truly riveting memoir."
    -Michele McPhee, Producer ABC News.
Best-Selling Author, Investigative Journalist,

"[My Father's War]...told in their own words--brings the heroism and heartbreak of World War II to life."
    -Wayne M. Barrett, Publisher and Editor in Chief, USA Today Magazine

"A fascinating compendium of personal narratives...the stories of WWII veterans from all ranks, all theaters, and all branches."
    -Blake Stilwell, USAF Veteran, Director of Communications and Managing

"You captured the experiences of their selfless, courageous, and patriotic service to America when it was needed most."
    -David Dion, Lt. Colonel, U.S. Air Force (Retired)

"You have captured the important stories of those who bravely wore our nations uniform."
    -Bob Dole, United States Senator, WWII Veteran

"Wear this cross, when you go to war you won't get killed or wounded. The Lord will protect you."

Virginia Thomas Arens

Dedication

To my Father
John Anthony Arens
My Hero, My Dad

To my Mother
Virginia Thomas Arens
My Rock, My Guardian Angel

Foreword

Its 10:00 AM Monday at the Military Heritage Museum in Punta Gorda, Florida. Captain John Arens (retired) is ready at his post as a museum guide. It could be any Monday as he is always in place, ready to start the day in his smartly pressed dress uniform, ready to greet the first visitors who arrive.

John is a storyteller just like so many of our volunteer veterans who serve as museum guides. For ten plus years John has shared his life story intertwined with the many stories that rise from the Military Heritage Museum's large collection of artifacts. As I walk the galleries, I spot John with a family of six huddled around him, capturing every word he says. He leaves them with a copy of a signed photograph of the USNS Redstone, the ship he skippered. It was the first thing he did when we first met – share his story on why he volunteers at the museum and then he gave me his signed photo.

"A Military Mustang" is the capstone to his life of service to his country. Service, as demonstrated by his commitment to volunteerism at the Museum, that keeps him going even into his 90's. With four decades of service across three branches of the military, John has lots of stories to share. Here at the Military Heritage Museum, so many visitors have heard his many stories of service and sacrifice. I am happy for him that so many others will now learn of his legacy and the impact he has made.

Gary Butler
Executive Director
Military Heritage Museum

Preface

To begin my story, I need to start with my family, before I was born. My grandparents, Paul and Barbara Arens, came to the United States in 1893 from Germany. My dad was born the next year on March 18 of 1894. He had five siblings, Arthur, Tony, Matthew, Mary and Ann. A little over 10 years later, my grandparents bought a new home on Mason Street in Hammond, Indiana where they lived for the rest of their lives.

When my dad grew up, he joined the Navy. He trained at the Naval Training Station in Great Lakes, Illinois. I am very proud of the fact that my father was in the US Navy during WWI. He served on two naval ships, the USS West Loquassuk and the USS America.

As a young boy, I looked up to my dad. He set a good example for me as I saw the things he did to help others. Some things were little, everyday kind of things. Others were life changing. For example, in 1925 my dad was working on the railroad. He was walking along the tracks when he heard somebody hollering under the train. My dad crouched down to try and determine where the sounds were coming from. He followed the sound till he came to the person who was doing all the hollering. It was one of his buddies. The guy said to my dad, "John, I'm pinned between the wheels and when the train moves it will cut off my head."

My dad ran up to the front of the train and told the engineer about his buddy being trapped between the wheels of the old train. Dad asked the engineer to move the train just one foot. The engineer told my dad to go back to his buddy and tell him that he'd do his best. "Notify everyone on the railroad," the engineer said, "that when I move the train, they must pick up the slack between the cars." The *slack* is the short few inches between cars. If used by a good engineer, the engine moves first, gradually adding the weight of one car as the 'slack' is taken up, without needing to pull all the cars immediately. My father got set to pull him out when the slack was taken up from the car in front of his buddy.

Anxiously, my father was waiting for the slack sound to get nearer. He waited until he heard the next car's slack sound. When it got to him, he quickly pulled his buddy out from under the track, saving him from a horrible death.

Years later, when my dad passed away, his buddy came to the funeral home. I remember, as I stood next to the casket his friend looked down at him and said to me, "Your dad saved my life when I was caught between the wheels of a train." I told him that my dad kept the old newspaper article in his trunk, and he had told me the story when I was just eight years old.

As my dad grew older, he started growing roses. When people came by, they would praise my dad, saying how beautiful the roses were. My dad also gave peanuts to the squirrels that would come right up to him and take the peanuts out of his hands. I think they knew it was easy pickings at the Arens' house. My father passed away on December 28, 1964. He was 70 years old. I relocated to Florida about that time. Shortly after I visited Dad's grave. Lo and behold, there was a squirrel sitting on his tombstone. I thought that was the squirrel's way of letting me know how much he enjoyed the peanuts my dad gave them. The squirrel on my dad's tombstone is one of my fondest memories. Through the years, I have tried to let my father's legacy of helping others inspire me to do what is right by my fellow man.

It is said, that a mother's love can't be replaced, and I would have to agree. I hope the story of my mother's love will outlive me through this book. I was preparing to leave home for the war, heading to New York where I would report for duty on my first ship, the ESSO Philadelphia. I remember it clearly. It was a sunny Tuesday morning, October 24, 1944. My mother was standing in front of me. She was holding a beautiful gold cross dangling from a gold chain. She reached up to put it around my neck. When I lowered my head to accept the gift she said, "When you go to war you won't get killed or wounded. The Lord will protect you." I wore the cross during the entire 37 years I served my country. In good times and in the darkest days, I was sustained as I could feel my mother's love and knew she was praying for me. I have been wearing that cross now for over seventy-years.

After leaving home, I never returned to live in the old homestead. I visited, but just for short periods of time when I was on leave from the service. Mom was born March 22, 1904 and passed away in Perrysburg, Ohio on October 19, 1971, she was only 67 years old. I believe that all her days, she was praying for me. Mom's love has sustained me through the years. I have held her close to my heart longer than the days she walked the earth.

Author's Note:

I first met Captain John Arens at the Southwest Florida Military Museum and Library located in Cape Coral, Florida in the summer of 2018. Even at ninety-one years old, John stood tall and proud in his highly decorated Merchant Marine dress whites. His wife, Lois, stood smiling and beaming at his side.

John's background is amazing. He served his country during WWII, the Korean War, the Vietnam War and was even called out of retirement during the Gulf War. John's military service of thirty-seven year's starts when he was rejected as a recruit from the US Marines—at seventeen; he was underage. So, he joined the Merchant Marines to serve during WWII. Interestingly enough, he was drafted by the United States Army in 1950 after serving in the Merchant Marines and getting a US Coast Guard discharge. From there he went on to the United States Navy. During the Navy years, John became a Navy Seal, an Arctic Diver, an Army Ranger, a paratrooper and a demolitions expert.

John's story starts as an impressionable but tough kid and continues as he grew into a disciplined leader. John Williams Arens is known as a *Mustang*. In military terms, that's slang for an enlisted man who becomes a commissioned officer without a formal college degree and military academy training. Although rare, it is not unheard of. Some notable *Mustangs* were Chuck Yeager, Audie Murphy, James Mattis and Wesley Fox.

The term *Mustang* comes from the horse called a mustang. The mustang is a feral horse that was able to be trained. The mustang was considered smart and calculating because it lived in the wild. However, caution was needed when working with mustangs. At times the old ways came back and overpowered the mustang's newer, more formal training.

The world of a Military Mustang is a world where very few ever venture. It requires a level of service to our country that most "lifers" never attained. It was a world where becoming a Navy Seal was just a beginning. Becoming a paratrooper was not enough. Deep-freeze diving was just a part of the plan. No, becoming a *Mustang* requires going above and beyond the call of duty.

Charley Valera
Award-Winning Author, My Father's War:
Memories from Our Honored WWII Soldiers

Acknowledgement

To all who served.
Thank you for your sacrifices.

Map of Ship Routes

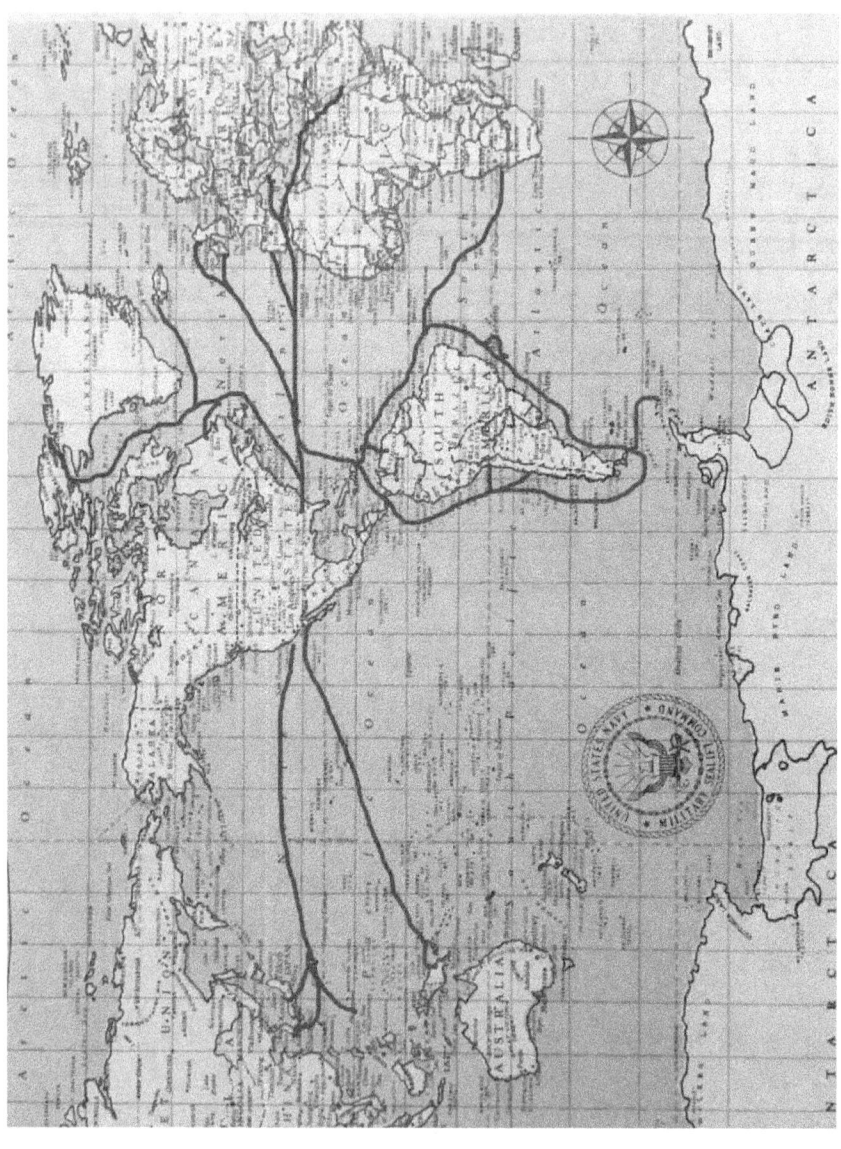

Contents

INTRODUCTION 1
Chapter 1 – SCHOOL YEARS 3
Chapter 2 – RELATIVES 11
Chapter 3 – FAVORITE MEMORIES 17
Chapter 4 – TOO YOUNG 23
Chapter 5 – MERCHANT MARINES 27
Chapter 6 – GOOD TIMES 35
Chapter 7 – TROUBLE ON THE WATER 41
Chapter 8 – UNITED STATES ARMY 49
Chapter 9 – NEAR MISS 59
Chapter 10 – LIFE IN THE ARMY 63
Chapter 11 – HELPING OTHERS 71
Chapter 12 – DAMN 38th PARALLEL 77
Chapter 13 – MYSTERIOUS WAYS 85
Chapter 14 – CIVILATION LIFE 89
Chapter 15 – NAVY DIVER 99
Chapter 16 – ARCTIC ACTIVITIES 111
Chapter 17 – NAVY DIVING TEAM 119
Chapter 18 – RECOVERY 127
Chapter 19 – DANGERS OF ARCTIC DIVING 135
Chapter 20 – EVERYDAY OPERATIONS 147
Chapter 21 – TRYING TIMES 157
Chapter 22 – MANY DUTIES 163

Chapter 23 – PROBLEMS 169
Chapter 24 – PROTOCOL & PROCEDURES 175
Chapter 25 – FAMOUS MEN & WOMEN 183
Chapter 26 – HELPING HAND 193
Chapter 27 – SHIPS 197
Chapter 28 – CHALLENGES AND AWARDS 201
Chapter 29 – OBSTACLES 211
Chapter 30 – MIDDLE EAST 219
Chapter 31 – TWILGHT YEARS 225
NAUTICAL GLOSSARY 229

A Military Mustang

INTRODUCTION

Welcome to an inside view of a *Military Mustang*. My story is about going from an enlisted man with a modest family background to a Captain in the United States Navy. Unlike most officers that achieve a high rank, I didn't attend the usual route of completing college and going to an accredited military academy. I did as I was ordered, followed my heart, and prayed I was making the right decisions during the many times I was involved in a world crisis. I served my country for 37 years, from 1943 to 1991.

None of us really know what lies ahead for us. What kind of "Big Picture" is in store. As kids growing up after WWI and during the depression, we learned to enjoy the little things in life. If you were lucky, a movie here and there, maybe a bike to get around on, and a great sense of family. I'd guess some of those values hold true even today. I'd like to think so.

As we all know, even the best laid plans can change in an instant. Looking at an old photo of me as a young boy, I realized that I would have never guessed the military life I would lead. When WWII broke out, I tried to join the Marines—as a 15-year-old kid I was quickly identified as too young and sent home. A couple years later, at 17, I decided my only option was to join the Merchant Marines.

Merchant Marines were supposed to transport cargo and passengers, but this was wartime. The Merchant Marines were an auxiliary branch of the United States Navy, carting fuel, ammo, and soldiers across the dangerous North Atlantic waters. As a result, we suffered a per capita casualty rate even greater than those of the US Armed Forces. According to the USMM.org, 1 in 26 Merchant Marines or 3.9% were killed during WWII. Casualty rates for other branches of service were lower. The Marines had 1 in 34, the Army had 1 in 48, the Navy had 1 in 114, and the Coast Guard had 1 in 421.

I hope that within these pages, you might understand my devotion to this great country of ours. As well as how respectfully I've been treated in return.

God Bless,
John

A Military Mustang

Chapter 1 – SCHOOL YEARS

As a young soldier fighting in Korea, my life experiences would escalate beyond a normal nightmare, into a world known only by being on the front lines of a war. The dangers in my life as a Merchant Marine appeared to be only a prelude to the horrors I would experience in combat, I just didn't realize it at the time. At the front, I was wondering if even Ma's cross and blessings would be enough to get me through this hell.

Gruesome Discovery

We were up on line in 1952 and I was sitting next to my foxhole. I looked around and saw a rock-like object but as I looked closer, I saw that it had hair on it. I gently dug around it with my bayonet and saw his ears. A few guys from the squad came over to help and we carefully dug away the dirt to expose the corpse. His fingers were shriveled but his gold wedding band was shining brightly. He was a 2nd Lieutenant who had just been made a 1st Lieutenant the day before. As I checked out the area, I saw a combat boot with his foot still in it. I hid it under some rocks to protect it from the elements while we waited on graves registration to arrive.

When graves registration came, we showed them the lieutenant. Mortar rounds had hit so close when he was in his foxhole that they killed and buried him. I told them how I had discovered him. He said, "Bring the boot to me. Hopefully, we will be able to read the name that should be written on the inside of the boot." I took the boot to him and they sent us back to camp. We walked back in silence as we processed what had happened.

Standing Up for Myself

I, Captain John William Arens, was born in Toledo, Ohio, on December 28, 1926 to John Anthony and Virginia (Thomas). My father's family was from Germany and my mother was of French descent. I have three siblings, Delores, Barbara, and Thomas.

When I look back on my life, in some ways it seems as I was preparing for a life of combat. Maybe I was destined for the military, to be a soldier and fight for my county. I think it all started when I was a 5-year old in the first grade at the neighborhood Catholic school in Hammond, Indiana.

On my first day of school, one of the boys said good morning in Polish (*jen do bra*). I didn't know what he said, and I thought he was making fun of me. So, I hit him! I got whipped pretty good as that was my first fight. Times were different than today. Everyone expected you to stand up to protect your honor. My father asked me only one question, "Did you fight back?" I said, "Yes." My dad said to mom, "Patch him up and sew his shirt up." The fighting continued, but I was learning fast. I started winning and began to like fighting. Finally, they left me alone.

Later that year we moved back to Toledo where I was enrolled in St. Thomas School. The first day the nun introduced me to the class. Charlie Shultz hollered from the back of the room, "BANANA NOSE!" (My nose was big.) It seems my body was not growing fast enough to keep up with my nose. I proceeded to the back of the room. When I got to Charlie, he made a big mistake. He stood up. I swung and decked him with my first punch. The Sister came running up to us, grabbed me, and turned me around. I said, "Why did you do that, Sister? Didn't you hear what he said?" She said, "That doesn't give you the right to hit him." I answered, "It does where I came from!" I got into fights every day at this school, too. The only difference was I started out winning the fights.

My Friend Harry

I'd like to say those fights were the only times I ever got into trouble, but my willingness to take risks seems to have caused me to get into trouble often. I spent many summers at Grandma's house. I cut the grass using an old-fashioned manual lawn mower with blades that would spin when it was pushed. I would trim the bushes in front of her house and do other things around the yard to help out.

Harold Grimmer lived across the street. We spent summers together because in my early years I was at my Grandma's house every summer. We had a lot of fun and got into a lot of trouble together. Harry was always coming up with crazy ideas. Harry's dad was often frustrated with our escapades. He was a court reporter and used shorthand to record everything that was said.

One summer when I was around ten, Harry got one of his crazy ideas. First, we put a chair in the yard. Harry took a picture of me sitting in the chair. Then, I got up and he set the chair on fire. He took another picture without advancing the film to create a double exposure. The double exposure made it look like I was sitting in a burning chair. The problem was this was his dad's favorite chair that he used to read the paper and listen to the radio.

All hell broke loose when his dad found out we had burned up his favorite chair. To make matters worse, Harry made it sound like it was my idea. Mr. Grimmer told my Grandpa to keep me over on his side of the street away from the Grimmer house. Harry's dad had very little patience with us, because we always seemed to be getting into trouble.

Our next episode occurred when Harry had the great idea to turn the garden hose on full blast and stick it into the front yard. With the full pressure on, Harry kept pushing it farther into the ground until most of the hose had disappeared. It kept traveling under the front yard like a snake. After a while the only part of the hose sticking out was about five feet from the spigot. The rest of the hose was in the ground. When his dad came out to use the hose, he really flew off the handle. To make things worse, the neighbor down the street told Mr. Grimmer that there was no way he could ever dig up the hose without destroying his whole yard, just cut it off below the ground level. Reluctantly, Mr. Grimmer cut the hose. I believe that hose is still there to this day.

A Military Mustang

Newspapers and Shoes

The depression hit in October of 1929 and everything changed. At one point, 37% of the people were out of work. Gone were the carefree days of youth. At ten years of age, I started selling newspapers after school to help provide for my family. A newsie had to defend his turf or get shoved off the corner by another kid who needed to earn cash for the family. Weighing in was an everyday occurrence. You could not show fear, or you would be finished. By the time I was 12 I had my own corner. I again found myself fighting for my ground.

I sold the most newspapers of my career in 1939 when the submarine, Squalus, went down. It was sitting on the bottom of the sea, off the northeastern shore of the United States. The watertight doors were closed, but some men were lost. Many were saved by the diving bell that fit over the hatch. It held eight men at a time. I went through the streets hollering, "EXTRA!" I sold 180 papers that day at 3 cents apiece. I made a penny for each paper sold. As I sold papers about the submarine, I had no idea that I would become a Navy diver.

Shining shoes was another way of making money. You worked hard to give a good 10-cent shine because you could make 5 cents extra in tips if the guy really liked how well his shoes looked. One man who was about six foot four inches tall put his size sixteen foot on my box, "Give me a shine." So, after I finished the first shoe, I politely told him that it was 10 cents a shoe. He took one hard look at me and said, "Son, you shine both shoes for

10 cents, or I'm going to hide this shoe up your ass." I said, "Yes, sir." When I finished, he gave me a quarter tip and said, "Don't be a smart ass."

St. Thomas School

By 1934, the county was well into the depression. We were living in Toledo on Halstead Avenue. The whole block was in dire need as many of the men were out of work. We were called Depression Kids. But not everything was bad – we didn't have far to walk to get to school and we passed the candy store. We would cut through Navarre Park and go up the alley past a little store named The Green Store. The store had penny candy. Suckers were a penny and 7-mint leaves were a penny. We would spend a lot of time just looking at the candy trying to decide what we wanted. My favorite were the peppermints. They were 7 for a penny, too. I usually chose peppermints because I got more candy for my money. Even as a little kid, I was very aware of the value of money and how to get the most for my penny. During the depression, I don't think any of the kids were wasteful – we just didn't have the money to waste.

At the Catholic school, the nuns were very strict. Sit quietly, no eating in class, don't talk back. One day, I was caught with a prune in my mouth by one of the Sisters. She called me up to her desk in front of the whole class and said, "What are you eating?" I said, "Nothing, Sister." She told me to open my mouth and take out what I had in there. I took out a prune. She said, "You lied to me. You said you weren't eating anything." I said, "I wasn't eating it, I was just softening it up for lunch." She didn't think that was as funny as the class did. By the way, the prunes were so dry you really did have to soften them up. Even though she was mad, at least I didn't get into trouble for lying. The ability to verbally defend my actions came in very handy when I became Captain.

Every so often people on Relief went to the firehouse to collect a sack of white flour, a sack of prunes and a few other items. Relief was like welfare or public assistance – helping people put food in their kids' mouths. I remember one time my sister and I were in school eating our lunch. We each had two pieces of bread. I said to my sister, "What have you got today?" She said, "I have a ham and cheese sandwich with lettuce and mayo." I said, "Me too." So, we ate our two pieces of bread with nothing in between while we acted like it was the best sandwich we ever had. It was "Touch and Go," during the Great Depression – you never knew when you would have food and when you would just have to pretend you had a great sandwich. As a Captain, I often had the ability to help others. Hard times helped me empathize with others.

In the 6th grade, my teacher was Sister Carolyn. She didn't put up with any mischief from boys. The girls had no problems with Sister Carolyn because they were good girls and weren't always into mischief. I was a different story as I seemed to always be in trouble even though I was a server at mass. For some reason, at mass I always prided myself for kneeling at the altar, being very straight, and not fidgeting with my cassock that hung down to the floor. But in school, I just couldn't seem to follow the rules.

When I was good, Sister would call me, "John." If I was causing trouble she would say, "JOHN W. ARENS, YOU BOLD PIECE OF HUMANITY." Hearing that statement told me I was in deep trouble. Sister had me bend over the desk and grab the other side. Sister Carolyn would use a leather strap. If you moved your hands to grab your butt, she gave you an extra one.

Sister Carolyn voted me "Least likely to succeed." Years later, I went to the school where she was teaching across town. I was an officer and wore my white uniform with all my ribbons and other decorations. I first talked to the principal and told her that Sister Carolyn taught me in the 6th grade. She took me to her class and introduced me to her and the class saying, "Sister, this is John W. Arens; you used to teach him in the 6th grade." I really think she turned pale from shock.

I stood in front of the class and I praised Sister Carolyn for teaching me discipline and helping me to become what I am today. I told the students that I owed my success all to Sister Carolyn because to give orders on a ship you have to first learn to obey orders and not cause any problems on the ship. She was one very surprised nun. Before I left, I walked over to her and said, "Would I be able to give you a hug, I need it." I knew girls could give nuns hugs, but I never saw a boy do it. When I gave her a hug, I whispered in her ear, "Thank you." I am sure when she said her prayers she said, "Thank you, Lord. The lost sheep has come home."

Chapter 2 – RELATIVES

My Dad

I was three years old on October 29, 1929, when the stock market crashed, and the country was plunged into the Great Depression. It hit hard and the world changed for the whole country and for our family. My father, like many fathers in the United States, was laid off. One statistic says 37% of the workforce was laid off. Layoffs were degrading to fathers who could no longer provide even the basic necessities for their wife and children.

Everyone on our block was poor, the depression spared no one. However, as bad as it was, we kids didn't know we were poor because everyone was in the same boat. However, fathers all over the country lost their self-respect, some turning to alcohol. My father was one of them which was a terrible thing, seeing my own father lose his self-esteem. I condemned my father and hated him for it.

I sold newspapers and shined shoes, making more money than my dad and shared it with my mother. The reason I am telling about my dad is because you will read later in this book about my own drinking problem with alcohol. I kept the problem hidden from my own father and mother for ten years, leading them think I was a non-drinker.

Summers at Grandma's

Whenever I was at my Grandma's house I slept up in the attic like my dad and my Uncle Larry did when they were young. I used to climb out the window and go down on the roof in the front of the house. One day, I climbed out the window and went farther down to the roof over the porch. I slid down to the ground on one of the columns supporting the porch roof. It wasn't easy but I just wanted to know I could do it. Of course, this was nighttime, but the streetlight on the corner was bright enough that it was easy to see what I was doing. Fortunately, I never got caught.

I spent a lot of time swimming at the lake in the park. Even though I could swim like a fish, Grandma was always worried about me drowning. Grandma told me I couldn't go swimming alone because she was responsible for my well-being while I was at her house. One summer when I was about eleven, I left the house wearing my swimming trunks under my clothes. When I got to the park, of course I went straight to the swimming hole. After a short time, one of my friends ran up and said my grandmother was coming towards the park and she had a towel in her hands. Grandma was an expert with that towel. She would snap it out and it would hit me on the legs like a whip. I did not want to get hit by Grandma's towel!

I put my clothes on over my swim trunks as fast as I could and ran the other way so Grandma wouldn't see me. When I got near the park entrance, I climbed up a leafy tree and sat on a branch and waited. Grandma

A Military Mustang

walked right under me as she started out of the park. She looked back one more time and headed back down the street to her house at the other end of the block-long street.

I waited until she got to the house and then I started down Mason Street, for dinner. When I got there the food was on the table and ready to eat. My Grandma said to me, "Where have you been? I was down to the park and I didn't see you." I said, "I was at the softball diamond at the far end of the park." My Uncle Matthew who was sitting behind me said, "Ma, I think the kid was swimming because he is all wet. He must have his swimming suit on underneath his clothes." With that, Grandma started chasing me around the table with her towel. She never missed, so I hid under the table where she couldn't reach me. Later I heard my Grandpa say to Uncle Matthew, "You didn't have to tell on the kid!"

Theodore Roosevelt and Uncle Alexander

In 1933 I was six years old. I was standing with my family on Memorial Day at the entrance to Harrison Park on Mason Avenue. We were watching the big parade go by. At the front of the parade, there was a group of officers on horses. Right behind the officers was my great uncle on my mother's side of the family who was born when the family lived in Canada, Master Sergeant Alexander Beaufour.

When my great uncle passed in front of us on his beautiful horse, he took off his hat and brought it in a high arc down to his knee. Mostly to the women, but I will never forget that grand gesture. My mother explained to me, "That's your great uncle, Alexander. In 1898, when he was a young man, he fought with Teddy Roosevelt at San Juan Hill in Cuba." I loved hearing about my heritage, and I was so proud of my uncle.

Uncle Matthew

I was about six years old and staying with my Grandmother Barbara Thomas in Hammond, Indiana for the summer. My Uncle Matthew came home holding his shoulder. He told my Grandma that the front mirror of the bus hit him when he was getting on. He had actually been shot by the police. The next morning when I got up and was heading for the bathroom I passed by the front door. It was just getting daylight as I glanced out and saw a bunch of policemen with Tommy Guns and shotguns in our front yard. I assumed they were in the backyard, too.

I didn't know what to do, so I went to the kitchen and quickly started putting paper in the stove. At the time, we had a coal burning stove in the kitchen and Grandma would let me put the paper in the stove so she could light the fire. I added woodchips for Grandma. Grandma wouldn't let me start the fire because she didn't want me to get burned by the hot coal.

As I was putting the paper and woodchips in, I said to Grandma, "There are police all around the house and they got guns!" Uncle Matthew heard me talking to Grandma and went to the front door. I followed right behind him. He opened the door slightly and told them that the only other people in the house were his old mother and six-year old nephew. They yelled back, "Throw your gun out!" My uncle hollered back, "I don't have a gun. I will hold both arms above my head and walk out." When he did, they took him away. He had been involved with some other men in the robbery of a warehouse. He was later convicted and sent to Crown Point Prison. He was there at the same time as the infamous John Dillinger.

When my uncle finally got out of prison, he never got in trouble again. He learned two trades, bricklaying and electrical work, and worked at both trades for the rest of his life. When he was an old man, he went to live in the old-folks home. The home had a big fire. Sadly, my uncle was one of the people who died in the fire. Some of the survivors said he was helping others get out and stayed in too long. I was very proud of my uncle, that he was willing to give up his life to help others live. That made him a hero!

Matthew Thornton

Matthew Thornton from New Hampshire was one of the men who signed of the Declaration of Independence. My sister, Barbara Arens, married Donald Keller. Donald's mother was a Simmons. The Simmons are descendants of Matthew Thornton. This makes Barbara's children, Becky, Cindy, and Tommy, and her grandchildren (Christopher, Danny, Robby, and Shadi,) direct descendants of Matthew Thornton.

Barbara had another son by her first husband. His name was Douglas. When he was a little boy, I used to give him rides on my motorcycle. I put him in front of me so that he would be safer. When he grew up, he bought a beautiful yacht. He was an expert on motors and other boat owners would ask him to work on their boats. He kept them in good condition. Everyone thought he was a great guy.

Chapter 3 – FAVORITE MEMORIES

Radio Debut

In 1932, I was five years old living in Toledo, Ohio. There was a radio station, WSPD, that advertised donuts during *Uncle August's Kiddie Show.* They had little children singing songs. My song was, *Everybody Works in our House but My Ole Man.*

I was sitting on the piano while waiting my turn. I turned to look at someone in the radio station, slipped and fell off the piano. I hit the floor and began crying. I was still crying when it was my turn to sing. All my relatives were listening to their radios, even my maternal Grandmother. I started singing while I was still trying to hold back the tears and not cry. In between my lines I went SNIFF-SNIFF SNIFF all the way through my song.

The people in the studio thought it was cute, me trying not to cry, dressed in a little sailor suit with a white hat. Little did I know that I would wear a white suit when I grew up and became a captain in the Navy.

Grandma's Apple Pies

My Grandma made four apple pies every Sunday and put them on the back porch to cool. When I was nine, we were visiting Grandma's. While everyone was in the front room, my cousin, Arthur, and I sneaked out to the back porch. We saw the pies sitting on a table with a tablecloth. I grabbed one of the pies, and we ducked under the table to eat the pie, hidden by the long tablecloth. Between us, we ate the whole pie with our fingers. It delicious, but really messy.

In a little while, Grandma came out to see if the pies were cooled and saw one was missing. She rushed back to the front room, demanding to know what happened to the missing pie. Our mothers found out that Bugs and I had eaten the missing pie. We got a whipping, but it was worth it! Grandma wasn't too mad. She said, "If the boys like my pies that much, I will make five pies so that they can have one for themselves." Of course, we loved Grandma for saying that!

Grandpa Thomas

I loved going to my grandparents' house. Grandpa Thomas chewed tobacco and had a full head of silver hair. Grandpa was in his eighties, but his stomach still rippled with muscle. He loved to listen to baseball games with his head down near the radio. He knew the batting average of every player on both teams, no matter which teams were playing that day. He called me Johnny Johnson every time he saw me.

I remember one time when I was about nine years old, I was over at his house on Utah Street. The bathroom door was open, so I went in to watch Grandpa shave. He used a straight razor to shave every morning. He had a leather strop hanging in the bathroom next to the sink. He would take a towel and hang it over his shoulders with his suspenders holding up his pants and no shirt.

A Military Mustang

He had just finished shaving and was holding his shaving cup with soap in it for lathering his face. He looked at me and said, "Johnny Johnson, I think it's about time for you to start shaving." Excitedly, I said, "Really, Grandpa?" So, he lathered my face, took the straight razor, and started to run the blade down my cheek. He rinsed the foam off and started down my cheek again. Just then Grandma walked in. She started screaming at him for using a straight razor to shave me. Turns out, Grandma and I didn't know it, but Grandpa was pretending to shave me with the rounded back side of the razor. He had to tell Grandma so she would stop screaming. Grandpa was just "funning with me."

Grandpa Paul Arens

My Grandpa and Grandma came over to America in 1893 from Ault Germany. Ault means old in German. Grandpa Arens was a shoemaker. So, he brought some of his equipment with him when he came here and started a shop. He put soles and heels on the bottom of our shoes. His work was beautiful and very high quality. He was proud of his workmanship. His son, John was my father.

When I was a little boy, around six, I would help my Grandpa Arens make grape wine in his basement in a very old wooden press. We put the grapes inside a pillowcase, and then I helped him turn the press. The juice would drain into a bucket and we poured the juice into a large barrel. The wine barrels would be put under the front porch where they began to ferment, turning into wine. Grandpa told me that in the old country, when guests came to visit, they gave them a glass of homemade wine served in a beautiful long stem crystal goblet.

I remember Grandpa loved to go down to Harrison Park. He sat on a park bench with Civil War veterans. He came to America in 1893, so he had no stories about the Civil War, but he liked to listen to their stories which were always interesting. Grandpa hated to part with a nickel, so I knew the best time for me to ask him for a nickel was when he was on the bench with his buddies. I also knew that I was not supposed to ask him when he was with his friends. He carried a long purse. When I asked for a nickel, he would unfold the purse and open the clasp on top with his fingers, shake the coins to the front, take two fingers to reach down and pluck out a nickel. Once I had the nickel, I used it to buy an ice cream bar from the Good Humor Man. I loved the uniform the Good Humor Man wore - a white peaked hat, white shirt, black tie, and white pants. And his ice cream tasted soooo good! I think I associated his uniform with the ice cream.

One day, I was talking to a neighbor who lived down the street. He was the same age as my Dad. He told me, "Your Dad and I would get into trouble when we were young kids. Our fathers would whip us with their razor straps." Many fathers used the strap in the old days which saved their hands. I told him that I didn't believe it because my Grandpa was a kind old man, he would never hit his son with a strop. He said, "I'm talking about when he was a young man. We had a lot of fun, but sometimes we did things that got us into trouble and your grandpa would take the strap to your dad."

A Military Mustang

Putting on a Good Show

I had a black friend when I was young. Wayne and I used to meet at Navarre Park to play. The older white men would tell me, "If you beat-up the black kid, we will give you a nickel." (They used the "N" word all the time.) I said that I would fight him for two nickels. Wayne and I would give them a good show and then I would give him one of the nickels. Wayne and I felt it was a great scam and it was only right that we should both profit from the ruse. We both made money, but sometimes it would get rough if one of us hit the other too hard. When we were older, we were 187th Airborne Troopers together during the Korean War. We joked about making money fighting each other when we were kids.

During fourth grade at St Thomas, I had many fist fights in the alley behind the school and in Navarre Park. Fred Norris always met me at the park, and I had to fight him. He would take off his glasses at the drop of a hat to fight. Calling him "Four Eyes" was a big mistake. I never did but being the new boy in school I ended up fighting him a lot. We would get dirty and tear our shirts from rolling on the ground. After many fights we became friends. I told him that some day when we got older, I was going to whip his butt good.

Years later I ran into him. He was 6' 3." I had just come back from the Korean War where I was an army Ranger. He looked at me and said, "You still want to whip my butt?" I laughed and said, "You would win right from the start." I'm glad we could joke about it because

I was a Ranger and he was an insurance salesman; it would not have been a fair fight.

A Military Mustang

Chapter 4 – TOO YOUNG

The Railroad Bridge

My dad always told me two things: Do not jump on a moving freight train and do not swim in the river. The river by our house was the Maumee River which empties into Lake Erie. It is one of the few rivers in the United States that run north. He always said, "If you do either, I will beat you to an inch of your life!"

One day I was down at the river standing on the railroad bridge where the train trestle crossed the Maumee River. The framework went 70 feet above the tracks to provide support from the top of the bridge. Where I was standing was even with the top part of the railroad boxcars. My dad, who worked for the railroad, was standing on top of one of the cars as the train crossed the bridge. The men often stood on cars to make sure everything was ok or just to enjoy the scenery, especially as the train traveled through their hometown. This day, as the train passed me my dad was looking right at me. I didn't know what to do because I thought I was sure to get beaten when my dad got home. So, I turned and dove into the water 60 feet below. When I got home, my dad never said a word to me. Years later I asked him why I didn't get the licking he promised me. After all, he had looked right at me and there was no doubt that he saw me. He answered, "Yes, I saw you. I was going to give you a beating, but then I realized you must have done this before. The reason for the threat that I would beat you was so you wouldn't fall off the bridge and drown in the river when you were younger." After you dived, I knew you had to be a good swimmer now, so there was no need for the beating. This was the same bridge I later sailed under on the lake freighter the Pegasus.

Marines

In 1942, I was riding the train from Hammond, Indiana with my family to New York's Grand Central Station. We were on our way to visit my grandparents. My father worked for the railroad, so we were riding on his pass.

While on the train, I went to the rear of our car to go to the restroom. When I left the restroom, I looked back to the next car. I was bored just sitting on the train, so I thought maybe I would take a walk through the next car to pass the time. When I opened the door, I saw wounded Marines, all with bandages on their arms, legs, or heads. When I stepped inside the railroad car, the conductor stopped me and said I couldn't go in there. The closest Marine to the door said, "Let him come in." They made room for me to sit down and started talking to me. One of them said, "How would you like to be a Marine?" Of course, I said, "Yes." Little did I know they were playing around with me. I never told them how old I was, but they could tell I was too young. They said that during the Civil War young boys would put a piece of paper in one shoe with 18 written on it so when they were asked how old they were, they would say, "I'm over eighteen,

A Military Mustang

sir," meaning they were standing over 18 so they really wouldn't be lying. I didn't know if that was true or not. Later, I found out it was. The Civil War is sometimes called The Boys War because there were so many young boys fighting. I suppose this was part of the reason.

After visiting my grandparents, we went back home to Toledo where I worked at a shoe repair shop shining shoes. The owner was Nick Poulous. He had no time for shining shoes because he had a contract with the Army to repair soles and heels on Army boots. He worked 10 to 12-hour days. I asked him to put extra soles and heels on my shoes to make me look taller. Just before I went out the door, he put some black shoe polish under my nose to make a mustache. I think Nick was laughing behind my back as I went out with higher shoes and a black shoe polish mustache.

A short time later, I took the bus to downtown Toledo and went to where the Marines were recruiting men. I took a deep breath outside the door and then I walked in. I must have look ridiculous to the Marine at the desk, an old Gunny Sergeant, with me clumping up to him with thick soles. He took one look at me and said, "What do you want? How old are you, son?" I said, "I'm over 18, Sergeant." He said, "Take off your left shoe." Then he told me to take off my right shoe. He saw the paper with 18 written on it.

"You know they pulled that in the Civil War, son, you can't fool me with that. Who told you to do that?" I proudly said, "A bunch of wounded Marines on the train coming from California." He said, "Get the hell out! You go home to your mother and grow up." Then I said, "You won't tell my mother, will you?" He replied, "No, I won't, but I wish I had 20 more like you." Then he took out his handkerchief and gently wiped off the shoe polish mustache from my upper lip. I went clumping out the door feeling proud that I had tried even if I wasn't able to join the Marines.

Army Halftrack

They were looking for drivers for Army halftracks going to Bryan, Ohio, from Toledo, Ohio, in 1943. I didn't have a driver's license because I was not old enough, but I applied anyway. Nobody ever asked me if I had a driver's license, so they didn't know I was too young to drive. The vehicles were being moved in an Army convoy. All I had to do was follow the guy in front of me. The front had wheels and the back had treads, just like a tank. They made a lot of noise driving down the road, as the treads clattered loudly on the pavement. There was a 105 Howitzer behind the driver.

I had only one problem on the way. I didn't stay close enough. I turned on the wrong street and ended up in a neighborhood with people coming out on their porch to see what all the racket was. After a couple of blocks, I found my way back to the convoy. I thought it must have been quite a feat driving one in combat. Even though I had been unable to join the Marines, delivering a halftrack made feel as if I had made some contribution to the war effort.

Chapter 5 – MERCHANT MARINES

Mom Started My Career
My mother worked as a waitress in a restaurant in Toledo, Ohio near the Maumee River. The owner let us eat dinner on the weekends. On Saturdays and Sundays my job was to clear the tables and help my mother who also washed the dishes.

A freighter was docked for the winter since Lake Erie was frozen during the winter months. It would start sailing around the month of March. A seaman by the name of Iner Brodin was one of the men on the ship who ate lunch at the restaurant every day. My mother introduced Brodin to me and he offered to teach me how to splice wire and rope. I got a job on the freighter and this started my career on ships. My first ship was the Pegasus where I worked for five months. During that time, I also learned how to steer the ship. After five months, I decided to join the Merchant Marines.

Signing Up
The Merchant Marines was the only service that would take anyone at age 17, because it was considered a 'noncombat' job. In the other branches of the military, you had to be 17 ½. I asked my mother if she would sign for me to join the Merchant Marines. When she hesitated, I added that it was not fighting, and I would make $90.00

a month. I would send $50.00 a month for her and the kids. She replied, "Where do I sign?"

First Ship

So, I joined the Merchant Marines. I took the train to New York's Grand Central Station. Then I walked to the Seamen's Church Institute (called the Doghouse by Merchant Marine Seamen) to report for duty. The sleeping rooms were very narrow with just one bunk. Showers were down the hallway. I stayed there overnight but couldn't sleep very well because the EL (Elevated Commuter Train) made a loud screeching sound as it was making the turn to head back uptown.

The next day I reported to the ESSO tankers office which was a couple of blocks away near Manhattan Island. When I went inside, the first thing I noticed was the man behind the desk had tattoos up to his elbows. I told him that I had sailed the Great Lakes for Pickens and Mathers on a lake freighter named the SS Pegasus. I was taught by Chief Officer Iner Brodin from Norway how to steer the ship and how to splice wire and rope.

The man behind the desk was fascinated because he was from Norway. So, he moved me up a rank to able-bodied seaman, instead of ordinary seaman. My first ship, the ESSO SS Philadelphia was a tanker in Norfolk, Virginia, which was in a convoy heading for England. I reported to the ship with my black knit cap and sea bag. The first person I ran into going up the gangplank was a U.S. Navy Armguard Gunner, one of the 28 Navy gun crew aboard. The gunners manned 3"/50 on the bow,

A Military Mustang

5"/38 on the stern, and 20mm over the top of mid-ship house and after-house.

He asked what I was doing there, and I proudly answered that I volunteered. He said, "You must be crazy if you don't have to be here." I said, "But I do have to be here, you, dumb ass," meaning that I felt I had a responsibility to do my part to contribute to the war effort.

Pegasus Experience Pays Off

The seaman on the wheel did not know how to steer the ship. The pilot said to the captain, "Don't you have anyone who can steer this ship?" That's where I spoke up and said, "I just came from the Great Lakes and steered the lake freighter SS Pegasus, sir, and I can splice wire and rope." The captain gave me a dirty look and told the ship bosun to have me splice wire and rope tomorrow. The bosun and the captain were Norwegians.

The next day the bosun told the captain that I did a good job when I spliced wire and rope. He told him I learned from a Norwegian sailor. So, he was sure that my seamanship was excellent. The captain treated me with respect from then on.

Proud to Serve

I was proud to serve in World War II on the tankers. While most people think it was not a dangerous service because we were not fighting, the German U-Boats went after the tankers knowing they could be carrying up to 150,000 barrels of aviation grade fuel for bombers and fighter aircraft. We also carried a 28-gun Navy crew on board. So, while we weren't technically fighting or considered combat units, we were support for those fighting and we did have combat units manning the guns in case of attack. Fortunately, the ships I served on were never torpedoed.

Cram School

I continued to serve in the Merchant Marines after World War II. One day, two officers approached and asked me if I wanted to be an officer. I said, "No way." I had only completed two years of high school. I told them I could never be an officer because officers used a sextant to determine where the ship was. The officers said they would tutor me. For two years, I worked with the officers learning how to use a sextant. They worked the heck out of me! When the ship would dock, I wanted to go downtown on leave. But they would say, "No, we have work for you to do. Go to the chart room and correct charts."

Finally, in 1949 they said that they had done all they could do and that I was ready to go to a cram school. The Captain Allison School in Baltimore was a cram school near the Coast Guard. You could take your test for 3rd Officer there. I was at his school for three months when the instructor said, "Mr. Arens, you are ready to take your test." I said, "Do you really think so, sir?" He said, "I have been teaching men like you for a long time and I say you are ready." He was right, I took the test and passed.

A Military Mustang

Officer - Third Mate
The Coast Guard Commander who signed my license said, "I don't know what you are going to do for a job. They are tying up ships by the hundreds." Nevertheless, I got a third mate's job as soon as I got my license. Finally, I was able to go home on leave. At home, of course I boasted of being a 3rd Mate Officer in the Merchant Marines.

A Hero
One of my heroes was Captain Herbert Nelson. The crew members called him LORD NELSON, but not to his face. However, I do think he secretly enjoyed being called that as he admired Lord Horatio Nelson, the famous British Admiral who defeated the French Fleet. He lost his life in the Battle of Trafalgar when he was shot by French Marines high on the topmast. His big mistake was wearing his uniform with all his decorations while on deck giving the French a perfect target after his officers advised him not to dress like that. They brought his body back to England in a barrel of rum. He was given a hero's burial.

Secret Alcoholic
I would be dishonest if I didn't tell about my early life from 1944 to early 1954 as a secret alcoholic. My father was an alcoholic and as a young boy, I detested him for that. However, as a young man I quickly became one myself when I went into the United States Merchant

Marines. I have no excuse for that, but I would not have gotten away with it in the Navy. It was easier in the Merchant Marines because the rules were not as strict aboard merchant ships and there was less supervision.

I was on tankers during World War II crossing the Atlantic Ocean with Navy escorts. Sometimes there were as many as 60 ships. Even with the Navy escort, we were in danger from German U-Boats with orders to sink the tankers first. Many times, wolf packs would attack with as many as 6 or 8 U-Boats at a time. Admiral Döenitz was the German Commander in charge of all U-Boats in the German Navy.

I think the combination of being out to sea for long periods of time and the danger from German U-Boats contributed to my use of alcohol. As I began drinking more, I became very secretive, hiding whiskey in various locations on the ship. It was interesting because whenever I would go home on leave, I stopped drinking. I think that was because I didn't want to upset my mother and being there only a short time, I was able to resist drinking. I continued drinking for a few more years, until the Korean War started in 1950. At that time, I was drafted off the ship as an officer in the Army. In the Army, I continued to drink when I could get alcohol, but I was always careful to avoid drinking enough to get drunk. When we were actually in combat a combatant could not get booze.

After the Korean War, I went back on ESSO tankers and continued sailing a short time. In early 1954, I went to the officers on the tanker I was on and told them of my secret life of drinking the past ten years. They were surprised when I told them because they could not believe I could hide it so well. My fellow officers said they would help me, but I would have to do it, "COLD TURKEY." They told the captain about my problem and he asked me how many bottles of booze I had hidden on the ship. I told him I had 14 bottles hidden in different locations on the ship. He said, "Holy crap!" I quit cold turkey, surviving the withdrawal symptoms, and haven't had a drink since early 1954.

Chief Officer

Later in my career when I was a captain, I discovered I had a Chief Officer who was an alcoholic. Once when he had too much to drink, he barged into the Radio Officer's room with an axe in his hand. The Radio Officer panicked, but the Chief Officer said, "I am not after you. I'm after a mouse." After that incident, I knew I had to get him off the ship and into the drug program.

His wife asked me not to do this as she was sure it would ruin his career, but I told her that this would save his career. Two years later I was on a ship that was receiving a safety award from an Admiral. The captain of the vessel, which was the same officer, and his wife were there to receive the honor. When they saw me, his wife thanked me for saving her husband's job. I never told them I was an ex-alcoholic myself. I continued to help seamen get into recovery programs. Jack Lawrence or I would put them on my ship so I could help save their careers by making sure they got the help they needed to get off the booze.

Chapter 6 – GOOD TIMES

Clowning Around
In 1948, I was living in Baltimore going to cram school. Captain Ellison was running the school in the basement of the hotel. In three months, I learned all about navigation, rules of the sea, how to read charts, radar, and everything else a 3rd Officer in the Merchant Marines needed to know. I spent some time studying on Saturdays and Sundays, but there was still free time.

Every chance I could I spent at the Baltimore Pool, swimming and watching the clown divers. They noticed I was good at fancy diving. (I had been diving since I was thirteen and I was the anchorman for the swimming team in Toledo as a kid.) They asked me if I wanted to dive with them. I said sure, if they would teach me some of their dives. The first thing I had to learn to do was a back flip where I pretended to land on the diving board and then fall off. The clown diving suit didn't fit perfectly. It hung too far down my arms and was too long going below my ankles. I also had to do belly flops learning to bend at my waist just before hitting the water. Of course, pretending to fall off the board was a standard act. I thought I did pretty good for a couple of months.

My First and Last Motorcycle
In 1948, the ship was in port for an extended time getting repairs. So many of us went ashore quite often.

On one excursion, I bought a brand new red 1948 Harley Davidson from a man in Norfolk. His wife had given him an ultimatum: her or the bike! It had a lot of extras, including a seat for two. I bought it for $840; he just wanted to break even. As a bonus, he taught me to ride.

We went to a park that had a baseball diamond. I put the bike in first gear, and he walked alongside me while I got the feel of it. Soon, I was shifting gears as I rode. After an hour, he said I was ready to take it out on the street. I practiced for about 30 minutes on a street that did not have much traffic. Finally, I felt ready for a busy street. So, I took off - I felt confident handling the bike.

I dropped him off at his house and I headed back to the shipyard. Going through town, I saw one of my shipmates and asked him if he wanted to take a ride. He said, "Sure," and away we went. While riding on the bike and seeing all the sights in Norfolk, he said, "Gee, John. This sure is a beautiful motorcycle. How long have you had it?" I said, "About 5 hours. I just bought it from a guy whose wife told him that it was her or the bike." Then he asked, "How long have you been riding?" I replied, "About the same length of time – 5 hours." He asked, "You mean you are taking me for a ride, and you have only been riding 5 hours? Let me off this bike!" By this time, we were outside the city. I said, "The least I can do is take you back in town."

Jack Johnson

When I was a seaman, I had the privilege of meeting Jack Johnson in New York City at the 42nd &

A Military Mustang

Broadway theater. One day, I went in and I saw this huge black man sitting on stage by himself. I asked another guy who he was. The man answered, "That is Jack Johnson, he was Heavy Weight Champion of the World, in 1900." I was told that I could go on the stage and shake his hand.

When I approached the stage, I asked him, "Can I shake your hand?" He answered, "Come on up on the stage." When I got up in front of him, I put my hand out. I never saw a hand so large in my life. When I shook his hand, it seemed my hand was lost in the middle of his palm. His smile was as big as his hand.

He asked me what I was doing, and I told him I was on an oil tanker carrying 137,000 barrels of high-octane fuel for the fighters and bombers in Germany. He said, "You really have a dangerous job with the German U-Boats trying to sink you. I asked him only one question, "Do you have any words of wisdom for a young 17-year-old sailor?" He looked at me and quoted a few words. "This above all, to thine own self be true and thou canst be false to any man." Then he gave me a big smile and said, "SHAKESPEARE," to let me know that he was educated. He was killed in a car accident in 1946 when he ran into a tree.

Jack Dempsey

Later that same day, I was walking down Broadway and went by Jack Dempsey's Restaurant. I looked in the window and saw Jack Dempsey sitting at a table with some men. He looked right at me and motioned with his hand to come inside. When I did, he sat me right next to him and told me to order whatever I wanted. A waiter came over and said, "What would you like?" Jack Dempsey said to the waiter, "Give him a steak with all the trimmings." I thought I was in heaven. Two heavy weight champions in one day. Of course, I had on my sailor suit and hat. Jack Dempsey was almost as big as Jack Johnson.

Stunt Riding

In 1949, I made a ten-dollar bet that I could ride my motorcycle standing on the seat. I had to ride ten miles from Rossford to Perrysburg. Some friends followed in a car, just in case I got into an accident. I did a stunt on the bridge - I leaned over the back of my seat with my head on top of the back fender near the rear license plate. I was looking at the top span of the bridge to keep in my lane.

That same year I met a motorcycle stuntman who taught me some more tricks. He taught me to ride inside of a barrel with a trap door at the bottom. The most important part of the stunt was to start out slowly at the bottom. As your speed increased, you could get the bike to go up on the sides of the large barrel. Finally, you could go up and down on the wall. Coming off the wall was the hardest part. You had to slow down carefully – but not so much that you fell off the wall. You continued slowing gradually as you brought the bike down so you could stop. After doing it successfully, two or three times, I felt really good about the accomplishment. I never performed in front of a crowd. I just wanted to prove to myself I could do it. It was a fine Harley, but I sold it to a friend in 1950 just before going into the Army.

Pig Farmer

I was a 3rd Officer, in 1949 on an ESSO tanker. Some of the crew told Seaman Bob, who lived on a farm that I could read in the dark. I was in my stateroom with a book in my hand with the lights off. When he called me to go on watch, he knocked on the door, opened it, turned on the light and saw me with the book in my hand. He said, "The crew told me you could read in the dark. I didn't believe it; I do now that I have seen it for myself." I went along with the joke.

At a later time, Bob was talking with one of the crew members. The crew member asked him if he knew how to butcher pigs and he said, "Yes." He told him there were two pigs up in the forepeak and asked if he would go there to butcher one. Bob said that he would. The others told him when he got there to climb down the ladder and turn on the light. What he didn't know was that there were two seamen hiding in the dark forepeak. When he climbed down the two men grabbed his leg with tight fingers and made grunting sounds. Bob hollered, "They got me!" As he climbed up the ladder the men followed him a few feet and began pinching his legs and grunting like a pig. Even though the men were just trying to have a good time, when I found out, I made sure that was the last time the crew messed with him.

Sammy Lee - Olympic Gold Medal Winner

One day in 1947, while I was at a pool in Toledo, Ohio, I met Sammy Lee. He would win a Gold Medal for diving in the London Olympics the next year. I asked him for a few pointers. Sammy was very patient with me and I thanked him for his advice. He is still a hero to me.

Prime Minister Winston Churchill

My ship was anchored in Plymouth, England, the same time as a British fleet. Winston Churchill was on a passing Navy tugboat, giving what we thought was the "V" for victory sign with two fingers spread apart. What we didn't know was it wasn't our "V" for victory but a sign from the Battle of Cerci in France, 1140 in AD. The bowmen used the same two fingers to pull the arrows on their long bows in battle with the French. If bowmen were captured by the French, they cut off the two fingers of the right hand. If they won the battle, they would come home raising high the two fingers used to pull the bowstring with the arrow. Prime Minister Churchill was really saying: "We have our two fingers and we are going to beat the Germans no matter what it takes, through blood, sweat, and tears. Along with the will of the people, the help of the Americans, the free French and French General De Gaulle."

A Military Mustang

Chapter 7 – TROUBLE ON THE WATER

Gettysburg Torpedoed

I remember Alphonse Miller; he was in the 8th grade when I was in 4th at St. Thomas Aquinas. Alphonse joined the Merchant Marine Academy when he got older. During the summer months all academy men were assigned to ships between classes. Unfortunately, he was assigned to the ESSO Gettysburg the summer of 1943. They were off the coast of Charleston, South Carolina when it was struck by two torpedoes about 4 seconds apart on the port side.

The men abandoned ship as it burst into flames, sinking rapidly. There were 15 to 20 sharks circling the men in the water. Third Mate Victor Grescenzo saved the life of John S. Arnold by holding on to him as he swam to a lifeboat. Several other men joined him in the lifeboat. Others swam to the burned-out areas of the vessel thinking they would be safe there and the sharks would stay away from the burned-out hull. The Merchant Crew lost 37 men and saved 8. The Navy Gun Crew lost 20 men and saved 7.

The men were picked up by the SS George Washington. Alphonse lost his life in the incident, the year before I joined the ESSO Fleet. Alphonse went down with the Gettysburg, like many others. His tomb is at the bottom of the ocean without recognition. When I think of Alphonse Miller, I remember him as being a great drummer like Gene Krupa.

Texas City Explosion

Our ship, an ESSO tanker, was sailing towards Texas City, Texas. Our captain slowed the ship planning to arrive two hours later than scheduled. That saved all of our lives. I was on the wheel and I asked him, "Why did you slow down the ship?" He told me, "Because every time I come in, they tell me to slow down as they are not ready for us, so this time I decided I would slow down before they told me to." The French ship, Grandcamp was already docked. It exploded at the pier. If we had been on time, we would have been docked next to it. A total of 537 people died that day. It was the biggest explosion that ever happened.

Our ship was diverted to Bay Town, Texas, where we loaded our cargo. I thought of my mom and the cross I was wearing as once again my life was saved. I have a picture in my room next to my bed. It shows a young sailor steering a ship in stormy weather with rough seas. Christ is standing behind the young man with his arm outstretched pointing the way to safety. Next to the picture is a copy of the seaman's version of the 23rd Psalm.

Emergency Landing

In 1944, our ship the ESSO Philadelphia, a tanker, crossed the Atlantic Ocean in a convoy of 60 ships. As we proceeded through the Dover Straits, our crew looked

A Military Mustang

up and saw a B-17 bomber which was crossing our ship heading towards the White Cliffs of Dover. Two engines were out and a third one sounded like a washing machine. With just one engine still functioning properly, we knew right away that bomber was not going to make it over the cliffs. As the plane approached the cliffs, it crashed into the water just off the beach. Hitting tail first, the front of the plane hit last. And the miracles just started adding up. The first miracle was that it was calm day, so the pilot was able to make a safe landing, and the crew was still alive. The second miracle was that the plane had made it that far without a German ME109 seeing it and coming to shoot it down. The third miracle was that our ship was there to pick up survivors from the water. It was the only time I ever saw a bomber flying that low over the water. The Lord works in mysterious ways.

Missing the Boat

I was an able-bodied seaman on an ESSO tanker going up the San Juan River in Venezuela. About 30 miles upriver was the port. Inland about 3 miles was the town of Caripito. The ship docked to load a full cargo of oil. We were taking it to New Jersey where it would be refined.

While the ship was in port, I decided to go into the town of Caripito during the day. I planned to take the train back to the ship. Shortly after leaving the town to return to the ship, the train broke down and had steam coming out of the engine. The engineer told us to walk back to the ship. I walked down the tracks, but when I got to the port, the ship had already left and was headed down the river. This was a problem because soldiers who didn't get back to the ship, were put in jail until the embassy could take them to the ship.

I looked around and saw a fast Higgins boat tied up to the pier. The owner agreed to take me back to the ship. I jumped on board and it took about ½ hour to catch the ship. Since the tanker was loaded, I could easily step from the boat onto the pilot ladder that hung over the side of the tanker. When I jumped aboard, I looked up to the bridge and saw the captain motioning me to come up to the bridge. He said, "How did it happen? I replied, "No reason, just give me my punishment." He said, "There must be a reason." I answered, "You wouldn't believe my story." "I've heard a lot of them," he said. So, I told him about the engine losing all the steam and that I had to walk back. The captain looked at me and said, "That's a new one; I've never heard that before. It won't happen again, will it?" I answered, "No Sir, it won't happen again."

I never missed a ship again. Much later in my career as captain, one of my seamen was going to miss the ship. Remembering what it had been like to miss the ship, I waited till the last minute to leave. As I pulled away from the dock, I saw him running to the ship and I pulled back in.

A Sinking Ship

When I was still a seaman with ESSO during the war, I was told by an officer to report to a ship. I told him,

A Military Mustang

"I can't go on that particular ship since all my gear and clothes are on a different ship. I need to go back to the same ship I just got off." After a conversation lasting a couple of hours, they finally agreed to send me back to my old ship. That was my lucky day! The ship he was going to send me to sank in a storm off the South African coast. Remembering my mother's promise, I always thought my whole life at sea, in combat in Korea, diving in the Arctic, and even in hurricanes that someone above was watching over me.

Man Overboard

I remember an incident took place during World War II on a fully loaded T-2 tanker which has a 12-foot freeboard. Above the waterline we had a 12-foot steel deck with fighter planes sitting on top. They were covered to keep the saltwater from getting on the engines. We could carry 12 fighter planes on the afterdeck and 10 on the foredeck.

Seaman Lorenzo and I were walking on the starboard side forward of the wheelhouse, checking the deck cargo and the planes to see if the tiedowns were secure. It was a cloudy day, but no water was breaking over the deck. Lorenzo and I were about the same age. As we checked tiedowns, a freak wave broke over the side. Lorenzo was picked up and carried over the side of the ship. Tankers do not have outer bulkhead railings like cargo ships. They have only a small lip about 3 inches high so the water could run back into the sea.

Unfortunately, we were not wearing life jackets because the weather was nice, and the water was calm. This was considered acceptable by officers at that time. If the water had been rough, we would not have been allowed on the main deck without a life jacket. I immediately yelled, "MAN OVERBOARD! starboard side," to the bridge. Which meant make a 60-degree turn. As soon as the tanker hit 60 degrees, it would be brought back a hard left which would turn the tanker 180 degrees. Without this maneuver, the tanker would not be near the seaman. Since the tanker is like a large forklift, it turns from the back, not like a car with front wheel steering. The bigger the tanker, the wider the turn.

I kept the bridge informed of the location of Seaman Lorenzo by pointing to him. I could see his head and arms waving from where I was. The officers could see him with their binoculars. When the ship was far in its turn, I lost sight of him. I ran to the other side of the ship and saw him again. The officers immediately called the captain to the bridge. When the ship came alongside Lorenzo, they threw a life ring and pulled him aboard.

Don't forget, while all this was happening German U-boats were lurking off our coast during the war in 1941-1945. That's why we did not take the time to put a lifeboat in the water which would make us a sitting duck for Germans to torpedo. My buddy was so exhausted that he just flopped on the deck like a dead fish out of water.

Later, he told me he was exhausted because he had started swimming after the ship as soon as he hit the water. He thought I was somewhere in the water too and

A Military Mustang

feared I had probably drowned. So, he reasoned, if no one saw us go overboard then the bridge would not be notified of a man overboard.

He was happy when they told him that I did not get washed overboard and after I notified the bridge, I continued pointing to his location to let them know where he was at all times. He was surprised because he did not hear the whole story until we got back to New York.

Being of Italian descent, he lived in Manhattan in the Little Italy neighborhood for forty years. His mother and father had immigrated to the United States. His mama told him to bring me home with him so she could thank me for saving his life. She cooked a spaghetti dinner with all the trimmings. I had never been to an Italian household, so I did not know what a treat was in store for me. When we arrived, she met us at the door and gave her son a big hug and kiss on the cheek. Even though she was short, she was a rather large woman. She almost picked me off the floor, squeezing me and kissing both my cheeks. She said she was so grateful to me for saving her son's life. I told her it was nothing, I just hollered, "MAN OVERBOARD," and they turned the ship around and picked him up. She said, "But you were keeping him in sight and pointing to him so they could find him." She hugged me and thanked me again.

When it came time to eat, his mom filled my plate three times with spaghetti. I thought that was it. But then she cleared the table and brought out the meat and potatoes. Then they cleared the table again and we washed all of this down with homemade wine. Next, she served all kinds of fruit. I told my buddy, "I never ate like this before." He answered, "When we have guests, we always eat like this." His younger brothers and sisters helped serve the food and wash the dishes. I will never forget the kindness they showed to me that day. She must have blessed me ten times while crying the whole time.

Chapter 8 – UNITED STATES ARMY

Drafted

I was serving aboard the ESSO Greenville, in the Pacific in 1950. After we left Lima, Peru, we headed for the Panama Canal. I was told to report to the Captain's quarters. The Captain said, "I have good news and bad news for you." I said, "Give me the good news first, Captain." He replied, "You get to go home." I said, "I don't want to go home, sir. I just came aboard." "That's true, Mr. Arens, but you have been drafted into the Army." "But, Sir, I have been in World War II," I replied. "Yes. But being in the Merchant Marines does not count because the Merchant Marines is not considered a branch of the military. So, at the first American port I had to get off the ship and return to my hometown, Toledo.

After arriving home, I reported to the draft board. I gave the woman my name and she looked into my status and yelled as loud as she could in front of 100 people, "So you are a draft dodger! We have been looking for you!" Well, that really pissed me off. "Don't you dare call me a draft dodger! I was in the Merchant Marines serving on oil tankers during World War II. I made 3rd Officer in 1949, and I just left a tanker to report here. My ship was out in the Pacific Ocean and I had to wait for the ship to come to an American port before I could come home." She screamed back, "Don't you dare raise your voice to me; I'll have your ass. You report to the FBI in town; get over there right now! I'm calling them. You're in deep trouble!"

I took all my papers with me to the FBI office, including my letter from President Truman thanking me for my service and my papers from World War II. As well as papers from my current ship, the ESSO Greenville, that showed the date I boarded the ship and how long I was on the ship. The FBI agent said, "You are definitely not a draft dodger. Report back to the draft agent. Make sure you show her the dates and times you were aboard the ship. You do not have a problem." He added, "However, you have been drafted into the Army!"

I went back to the draft board the next day. The same lady tells me that my brother, Tom, has also been drafted. Now I am really pissed off. I told her, "Look, you bitch, you drafted my brother who is 3 years younger than I am by pulling his name from the back and moving him up. I'm telling you right now, if he gets wounded or killed and I am still alive, I'm coming back to this place and am going to hold you responsible. I will tear this office apart." She said, "I don't have to put up with your shit. You report back to the FBI office. You are going to jail!"

Back to the FBI office I went. The agent laughed and said, "What did you do now?" I told him what had taken place. I also told him that I called her a bitch for what she did to get my brother drafted. The agent said, "Look, she's got you and your brother by the balls. It won't help to antagonize her. Get out of my office and report to Camp Breckenridge on the date your papers tell you to be there. If you fail to report, we will step in."

A Military Mustang

I reported to Camp Breckinridge on November 1, 1950. I was almost 24 years old, six years older than the average 18-year old draftee. Some of the guys at Camp Breckinridge were 16 when I reported because the recruiting officer believed it when they said they were over 18. "God bless them."

Leadership School

I completed Basic Training with two of my buddies, Rocco Palombi and Stanley Peco. All three of us wanted to go to Leadership School. It was called "Army Forces Leadership Course." We attended from February 26, 1951, to April 20, 1951. This was an extraordinary experience and changed the course of our service.

When I completed my training, I went straight to Korea with my Basic Training Group as a combat unit. We later found out from some wounded soldiers that a massive group of Chinese had attacked their unit. Many were killed and wounded. One of the first men that was killed was a young soldier from a family that used to bring their car to the base with a trunk full of food on the weekends. The soldiers could choose whatever they wanted to eat. It was a treat to have 'real' food. It made us feel appreciated to know that a family cared enough to bring us treats regularly.

The soldier I was talking to said we were lucky. Because we went to Leadership School, we shipped out later and had not arrived when a massacre of another unit occurred. Saying a prayer for the lost lives, I again, was reminded of the cross and my mother's promise that I would be watched over by a higher power.

3rd Ranger Company

We became Army Airborne Rangers, 3rd Company on June 14, 1951. The commanding officer was Captain Jesse Tidwell. When we arrived in Korea, we took a train from Pusan north to the area south of the front lines and then traveled by truck to a Repo Depot just behind the front lines. We reported to a young officer in a tent. He told us to go to another tent with our gear and that we would probably not be able to stay together. We would be sent to the front lines one at a time, as he got word that someone was dead or wounded, one of us would replace him.

During our stay the 3 of us saw a soldier, who happened to be a sergeant from the 3rd Ranger Company, wearing a black beret with a bandage on his arm. He had a dirk knife on his web belt and a white pearl-handle 45 on his hip. I said to my Stanley and Rocco, "I wonder what the French are doing here?" He heard my comment and said in a southern accent, "I'm not French, I'm from Alabama. I have been wounded and am going to get my arm patched up. What are you guys doing here? You look too good with them new clean clothes on to have been here very long." We told him that we were draftees fresh from the States and we were at the Repo Depot waiting to be sent to the front lines when somebody got wounded or died. He answered, "You mean they are going to send you up to the front lines, wherever they need a body? You

A Military Mustang

won't last two days up there. Those men came up as a whole regiment, they don't want to know you and you will just be cannon fodder. How would you like to be a Ranger? We have dead and wounded up there. We need men now!" I replied, "We are not paratroopers. How could we be Rangers? He said, "Would all three of you jump without training?" We looked at each other and nodded, yes. Then we replied, "Yes, we will."

I asked him, "How are we going to do that? We have already been told to wait till we are sent to the front." He told us, "You go into the tent where you talked to the young officer and tell him you want to go up to the 3rd Rangers. They are right up ahead of where we are right now. Tell him you talked to me." So, we went back to the officer's tent and told him what the sergeant told us. The officer said, "I will have to figure out how to get you to the 3rd Rangers. If I can't, I can only keep you a couple of days then I will have to send you up as replacements. Do you understand?" We told him, "Yes, sir. Thank you, sir."

The very next day we were called to his tent. He said, "The Rangers are going to take you in, so get your rifles and I will have a jeep take you up as close as it can get to the unit. When you hear guns firing, the driver will drop you off. He has to come back here because the jeep belongs to me. Good luck!" Away we went, each of us with a canteen of water and our rifle.

As soon as we got close to the front we ran into a Ranger and we asked him where we could find Captain Jesse Tidwell. He told us that he was over to our left in a bunker. The Ranger looked like he just came off a patrol. He had a bandoleer of ammo strung all around his shoulders, a 45 on his hip, and a dirk knife on his web belt.

We arrived at the bunker and walked inside. Captain Tidwell met us and said, "Glad to have you men. What Airborne outfit are you from back in the States?" I spoke up first, a little confused. "We are not paratroopers, sir. We are draftees." About that time a Ranger sitting at a homemade desk with a small light, broke his pencil. I thought that we were in trouble as I realized the young officer never got in touch with the Rangers. Captain Tidwell said, "What the hell are you doing here?" I spoke up immediately and said, "Sir, you need men right now and we talked to the sergeant with his arm in a sling. He said you would take us if we would jump without training." Captain Tidwell knew who we were talking about. He took a long, hard look at me and said, "You won't make an ass out of me, will you? You will go out the door with no training?" I said, "Yes, sir, we all agreed." Captain Tidwell continued to look long and hard at me for about 30 seconds, pondering the situation. Finally, he said, "You go to Second Platoon with Sergeant Jenkins and you other two go over to Third Platoon." We gave him our traveling orders and left.

I reported to Sergeant Jenkins and he told me to report to Sergeant Elmer McCullough. When he found out I wasn't even Airborne, he gave me a disgusted look and asked me if my rifle was clean. I said, "Yes." He replied, "Get yourself some bandoleers of ammo. We are going out on patrol in the morning at 4:30. We want to be out there by daylight."

My hole buddy was Charles Murphy, an outstanding Ranger who was an expert with a knife. We went out on patrol and we were walking in a wadi. A wadi is usually full of water when it is raining in the mountains, but it was dry now. Sergeant McCullough pointed and hollered for me to walk up on the bank. I said to Charley, "What does he want me to do that for?" Charley answered, "So they will shoot at you. Then we will know where they are. Don't worry, we all took turns. Sergeant McCullough took the first turn. It is just your turn in the barrel."

I climbed up on the bank and started walking. Suddenly I saw dirt kicking up all around me. I did not hear the weapon fire until the bullets whizzed past me. Sergeant McCullough screamed, "Get down from there! We see them!" I jumped down and Charley said to me, "You're in!" That was my first experience in combat. Later, Charley spent a week training me how to use a knife.

One night, Charley said to me, "Do you want to be a real Ranger?" I said, "Yes." He replied, "I want you to leave our position, crawl or creep low and get to their position. You will know when you get close. You will hear them talking. Stay down, pick your man, stab him just like I taught you, and bring me back one of his ears so I know you did it."

Being an officer in the Merchant Marines I made sure I always had a compass with me. As soon as I got out in front of our foxhole, I buried three sticks in a row about 8 inches apart for a reference point. When I left, I took a reading on my compass and made sure I was heading north. I proceeded moving low, sometimes crawling. I was gone for 3 ½ hours when I heard voices ahead. At that point, I was crawling on my belly. I had made sure I had nothing that might rattle or make noise. I could see outlines of enemy soldiers up ahead. I picked out a soldier that was standing apart from the others. I was worried they could hear me since I could hear my heart pounding just like it was going to come through my shirt.

    I stood up slowly and made my move. He had his back to me, and I moved just like Charley taught me. I grabbed him around his neck from the back, pulled on his head as I sunk the knife in his neck and twisted it as hard as I could. The only sound was gurgling. I noticed my right sleeve was wet from his blood when I remembered that I had to cut off his ear. Once I realized he was dead, I crept away from him. I took a bearing on my compass to the south and got the hell out of there.

    Since I didn't run into anybody on the way up there, I felt I could stay in a low crouch rather than creeping so I could move, silently as fast as I could. It took me a while to find the 3 sticks I had laid in a row with a bearing of north on my compass. Now all I had to do was set the compass bearing south on the first stick and sure enough it faced south. I pulled out the sticks, scattered them as I moved closer to the camp. I whispered, "Charley, it's John." He said to come on in.

A Military Mustang

When I got next to him, he said, "Where the hell have you been?" I told him, "I did what you told me to." Charley said, "Holy shit! I thought you would crawl out in front of our fox hole and stay there for a while and then come back in and tell me it was too dark to find the enemy."

As soon as it started to get lighter, I showed him the ear I had in my pocket and my bloody right sleeve where the soldier's blood ran down from his neck. He said, "Hurry up, bury the ear! Captain Tidwell will kill me if he finds out what I told you to do alone as green as you are!" I promised him that I would not tell anyone.

187th Airborne Rangers

Charley was in the Vietnam War in the 60's and highly decorated. He has been dead a few years now, but he kept me alive while we were foxhole buddies. When they broke up the 3rd Ranger Company we went to Japan with the 187th Regimental Combat Team where I went to G Company. Charles Murphy, Rocco Palombi, and Stanley Peco went to H Company next to us in Beppu, Japan, training for combat that would come later. We got to see each other the whole time we were there.

I made sergeant in the 187th Regimental Combat Team in 1952. The unit went back into combat again in the same area as the 3rd Rangers. When our unit was back fighting again, I killed three more enemy soldiers with the knife, the same way as I did that first time Charley told me to. Only now I was more confident. I no longer had a pounding heart and I was sure of my skill. It is not something to be proud of. Since it was done in the nighttime when it was very dark, I never saw their faces. They didn't seem real and I didn't feel haunted if I never saw their faces. I did not let my men know what I was doing and never bragged about it.

Cabu Prison

During World War II, Camp Breckenridge had interred German prisoners of war. They lived like kings. When our men were held prisoners by the Germans, they were treated like shit; they were always hungry, and they were shot if they tried to escape. The Germans never tried to escape. Life was better for them in the American POW camp than back home. As bad as it was at the German camps for Americans, the Japanese treated our men much worse. They burned our prisoners alive in one camp. They were preparing to do the same at another camp when the 6th Ranger Battalion made a daring raid at Cabu Prison (Cabanatuan Camp) in the Philippines on January 28, 1945. The Rangers killed every Japanese guard before daybreak. Nearby there were 8,000 battle-hardened Japanese troops in the area, but they were not close enough to save their comrades. There were over a hundred Philippine scouts with the Rangers, covering their flanks. The book "Ghost Soldiers," by Hampton Sides tells about this event. Later, a movie was made about it called, "The Great Raid."

Chapter 9 – NEAR MISS

Deuce and A Half

In Korea, I was in the last deuce and a half truck in a convoy. A deuce and a half is a 2 ½ ton cargo truck, often used to transport troops. As we rode through a small town, a lot of Koreans were on the sides of the road. Generally, there was a good distance between each truck. Our driver was keeping a safe space behind the other trucks when a small toddler wandered out in the middle of the road directly in front of him. From the of back of the truck we couldn't see what was happening. The driver swerved as he hit the brakes causing the truck slide sideways.

Oblivious to the danger, the child sat down in the middle of the road and the truck slid right over her head without hitting her. When the truck stopped, we all hopped out to see what was happening up front. It was chaos, as everyone was shouting and praising the driver for missing the little girl. Unaware that it was a miracle she was alive, the little girl continued sitting in the middle of the road. Sobbing, her mother rushed out and grabbed her daughter, pulling her into an embrace. With all the death and dying of the war, it lifted our spirits to see the life of a child spared by the quick response of the truck driver. Everyone jumped back in the trucks and we continued on our way.

Breathing Underwater

In the summer of 1951, I was on patrol during a combat mission as a Private with the 3rd Ranger Company. We were in a big, wide water filled valley which was full of rice paddies. We walked on the dikes which ran through the middle of the paddies to keep our feet dry. Our Point Man who was ahead of the patrol, sent word back to us that a large group of the enemy was coming right at us walking on the dikes just like we were.

There were some hollow reeds growing in the area. Captain told us to break off the reeds, lay down in the water on our backs at the deeper end. Put the reed in our mouths and breathe through the reed. As quickly and quietly as possible, we jumped in the water. The enemy platoon passed by without seeing us. All the while, everyone was hoping no one panicked because they were breathing through a reed. This was just another one of my close calls as an Army Ranger.

Our Last Patrol, Ranger Ronald Racine

The 3rd Ranger Company was on its last patrol. That night we crossed the waterways and spread out, looking for signs of the enemy. When we got to land, we followed each other, single file. Whenever we stopped, we would put out our hand and touch the man in front of us to stay together. When we moved again, we stayed as close as possible. As we proceeded a little further up the trail, one of the men had his hand on what he thought was the man ahead of him, but he was really a little off to the side. It was a dead Chinese soldier. He didn't smell yet

A Military Mustang

which meant the person who killed him could have been our lead scout. So, our guy who touched the dead body had to catch up because the man in front of him had kept going.

As we went farther, we had to bunch a lot closer. We had a lead scout and an inner scout. Ranger Ronald Racine was the outer flank guard and I was the inner flank guard. We were moving slowly along when all of a sudden, guns started firing off to our left, about 45 degrees. Up ahead, I heard Ranger Racine holler, "I'm hit!" I moved up close to him. I felt blood on his stomach and his back. Quickly, I yelled for the Medic.

The Medic told me to keep up with the patrol and someone else would take care of Ranger Racine. I continued on with the patrol. We could still hear the enemy. We attacked screaming like crazy men. The enemy heard all of us hollering at the top of our lungs and hauled ass away from us because they thought they were being attacked by a whole battalion, not a few soldiers on patrol.

I didn't know how badly Ranger Racine was injured. I just knew he was taken to an aid station. He was then sent back to the United States to Waltham General Hospital in Massachusetts. I did hear that he was in a wheelchair for the rest of his life.

He passed away in 1988. I will always remember him as Ranger Racine, a young Ranger full of life. It wasn't easy to be a Ranger, and everyone knew that you had to be a real man to get through the Ranger Course at Fort Benning, Georgia. That he was and more as he served our country. He was a Ranger to the end, giving everything for his country.

A Military Mustang

Chapter 10 – LIFE IN THE ARMY

Brothers

When we were in combat, we used to carry black shoe polish during patrol. We put this on our cheeks and chin for camouflage. The black paratroopers used to kid us and say, "We don't have to use that shoe polish." I would tell them, "But your white eyeballs give you away in the dark, so keep your eyes closed as much as possible." We joked around with each other because it helped relieve the stress of being in constant danger. But no matter how hard a time we gave each other; we would protect each other even if we had to die to do it. We were brothers, period.

Airborne Troopers

Captain Tidwell was in command of the 3rd Ranger Company. He told the three of us who joined without training, to report to an airfield to make the jumps needed to qualify to be Airborne Troopers and get our wings. We had already engaged in combat as soldiers with the 3rd Rangers. So, Rocco Palombi, Stanley Peco and I were going to make the three parachute jumps to qualify to get our wings. We were heading for our aircraft, a C-46 with our parachute, our reserve parachute, and our pack which we had to carried between our legs. We were walking like pregnant women because of all the gear we had on. A bunch of Marines were leaning against their packs while they waited to get on another plane. One Marine hollered, "What's that on your backs?" Another Marine hollered, "Those are parachutes." A third Marine responded, "I thought they jumped without them."

Then one of our Rangers hollered, "How many men in a Marine squad?" A Marine replied, "12." A Ranger retorted, "One cameraman and 11 film bearers for their exploits." The Ranger Sergeant and the Marine Sergeant made their units do pushups for punishment due to speaking without permission. It was all in fun. At war, men often resort to humor to alleviate the stress of war.

When we reported to the aircraft, 2nd Ranger Company was there to make the jumps required every so often to get jump pay. Lieutenant Queen was in charge and told us to put on our chutes. I told him we didn't know how to put the chutes on because we weren't Airborne yet and had no training. He picked out a 2nd Ranger named Winston Jackson Sr. and told him to put chutes on us.

The Rangers told us, "You white boys must be crazy for jumping out the door with no training." That didn't stop us, we made our three jumps and qualified. In the field, we only needed 3 jumps to qualify. If we were back home in the States, we would have had to make five jumps at Fort Benning, Georgia.

Being a paratrooper got me more respect from the 3rd Company men. Major James Queen became a life-long friend. He sent me a picture of himself and wrote on the bottom, "Your first jump master." I treasured that photo and was happy that he thought enough of me to send it to me.

The Buffalo Rangers

The 2nd Ranger Company was an all-black unit with black officers. At this time, the military was still segregated. The 2nd Ranger Company arrived in Korea in the wintertime. So, their fighting was done in hard times due to the weather. They also had more casualties than all the other Ranger companies.

The 2nd Ranger Company called themselves, "Buffalo Rangers," named after the Buffalo Soldiers who protected settlers after the Civil War around 1870-1880. Legend has it that the American Indians had given the soldiers the name because of their black hair that resembled the hair between the buffalo's horns. While some may think it offensive, at the time it was an honor to be called Buffalo Rangers because the buffalo were honored by the American Indians. In Korea, the Officers took the name because of the honor it conveyed.

Years later after retiring from the military, Major Queen and I became best friends when we reconnected at a Fort Benning get-together in 1999.

After the Korean War a lot of the Black Rangers called each other for support. Some lived in the South and were treated as third-rate citizens. Others moved to Washington, D.C. where they were not treated as badly.

Some became mailmen because the U.S. Postal Service went by point system. If you did a good job, you got points, it didn't matter what color you were. Not only did those Rangers become mailmen, but their children were able to receive a good education, including the opportunity to go to college which allowed them to have a much better life.

The 2nd Rangers were proud to serve in the war. Sad to say, some of the black soldiers in the United States were sent to fight fires in Oregon instead of going with their unit and being allowed to fight in the war. At the time that caused a lot shame because they had trained for combat but were not allowed to fulfill that role. To make things worse, when the soldiers were in town with their jump boots on, white soldiers would cut off their pants at the top of their boots to disgrace them even more. The white soldiers didn't believe that they would jump out of planes. What a joke! These men were heroes.

Thanks to General Slim Jim Gavin who allowed them to march with the combat troops when the 82nd Airborne came back. I was proud to know these American heroes. All of Harlem came out to see them march with the white paratroopers. They were at the rear, but they were ok with that. They were doing their own thing and showed up the white guys in front of them.

All the paratroopers in front of them were walking together straight ahead in standard military formation, but the black paratroopers in the back put on a real show doing right oblique, left oblique, to the rear march. They brought their rifles to their waist, flipping them around in different positions, rather than just the standard march.

I owe my life to the fact that some black soldiers were allowed to fight. I was saved by a black paratrooper during the Kojedo Operation. If not for the paratrooper, I would have been stabbed in the back in the midst of 100,000 prisoners that were rioting. Because it was in the middle of an operation, sadly, I was not able to learn his name. I was hoping someday he would come forward. Maybe if he reads this book, he will see the story and come forward. I am proud to have made my first jump from an airplane on July 27, 1951, with the Black Rangers and my two white buddies without any training, Rocco Palombi and Stanley Peco. HO YA!

Rockkasan

The 3rd Rangers broke up in early October. Being as we were Caucasians, we couldn't work behind enemy lines without being identified. We were transferred to the 187th Regimental Combat Team, known as Rockkasan, to train for another Airborne drop somewhere in Korea.

The 187th Airborne Regimental Combat Team was stationed in Beppu, Japan. We were riding in trucks when an officer stopped the trucks and said, "Everyone off the trucks! We are going to march the rest of the way to the camp like Rangers! We are going to march to a cadence, and you are going to make me proud of you!" We gave them a real show at camp.

After our arrival, they split us into different companies, before sending us back to fight in Korea. I went with six others to G Company. The commanding officer was Captain Jonas Epps from Macon, Georgia. He had five silver stars on his chest. I was Private First Class when we arrived. Shortly after, I made Corporal and wore two stripes on my sleeve. I was also an assistant squad leader.

Respect but Verify

I was Corporal of the Guard at the main gate with orders not to let anyone pass without identification. One day, General Trapnell was coming through the gate in a staff car with a driver. The procedure was that Generals did not have to go through the same process as everyone else and did not have to stop at the gate, being as they should be recognized.

I couldn't see the General who was sitting in the back seat. I noticed his driver wasn't going to stop so I stood my ground in front of the car at port arms. When the driver realized I was not going to move he hit the brakes. That threw the General forward and his head hit the back of the driver's seat. (There were no seat belts at that time.) The driver started hollering at me for holding up the General's car.

I walked around to the side of the car and the General, having recovered from the sudden stop, rolled down the window. I spoke very quickly before the General could reprimand me, "Corporal Arens, sir, 52-008-276. I saw your star on the front bumper, but I did not know who was in here until I looked at you through the window. I have orders to stop everyone at the gate." The General said, "What is your Company and your Company Commander's name?" I answered, "G Company, Captain Jonas Epps, commanding, sir." He closed the window and ordered his driver to go.

A Military Mustang

The next morning, I was called into our Company Commander's Office. He wanted to know what happened. I explained, repeating everything that was said, verbatim. He said, "I wondered. The General told me to make you a Staff Sergeant right now, this very day. You did the right thing and we need more non-coms like you," he said with a smile on his face. I said, "Thank you, sir." He said, "Get the Staff Sergeant's patch and sew it on now. I'm moving you to a squad of your own. He also noticed your 3rd Ranger patch. Dismissed." I left his office and went to the Base PX, got my patch, and they sewed it on my sleeve. Then they took a photograph of me wearing the new patch.

Bataan Death March

We had the privilege of serving under General Thomas Trapnell. Years earlier he had been captured in Corregidor, Philippines, in World War II and had been on the Bataan Death March for 69 miles. If you were tired and fell to the ground, the Japanese soldiers would stab you with their bayonets. If another soldier was ready to fall down and you tried to hold him up and you both got too tired and fell, they would bayonet both of you. Thousands of prisoners died on the march.

Those who survived the march faced horrifying conditions when they were put into prison camps where they were literally worked to death. Soldiers at the camp were deprived of food, water, and medical treatment. Many of the camps had white crosses that were made of whatever scraps could be found to make a cross.

Approximately 10,000 men died on the march through the Philippine's scorching jungle trails.

General Trapnell, a Major at the time, survived and eventually ended up General in charge of the 187th at Beppu. The men who served under him had nothing but respect for the General.

Chapter 11 – HELPING OTHERS

The Garden of Light Orphanage

Some of my fondest memories of my years in the military was the time I spent at The Garden of Light Orphanage. The orphanage was near the base and I spent a lot of my off time working with Sister Mafalda, who ran the orphanage for Japanese children. The sisters from the orphanage took the children and hid in caves to protect the them.

At Easter time in 1952, I was teaching the Japanese orphans how to sing sacred Easter songs in English. I would say the word in English and a nun would write it in Japanese. It worked! By Easter the children knew five songs in English. On Easter morning after the troops were seated, the children were brought in and seated in the choir loft. When they started to sing, everyone was surprised because there were no English children on the base, yet I was directing a children's choir that was singing English Easter songs. I was so proud of my little choir and all the work we had put into learning the songs. The little choir was the highlight of the service.

Sister Mafalda was actually from Italy and she was blessed with a vision from the blessed Mother of God, Mary, Mother of Jesus Christ. A few years before this, she was very sick and at death's door. She was lying in bed, too sick to get up and Mary came to her and stood at the foot of the bed. While the other nuns were at Vespers (prayers), Mary said to her, "Get up out of your bed, get dressed, and go to church because you have much work to do."

She got dressed and went over to the church and walked inside; all the nuns were flabbergasted when she walked in. They couldn't believe their eyes; they had just left her in bed, too weak to get up and get dressed. She told everyone about her vision and what happened. Sister Mafalda told me she had never been sick since that time.

She always called me, "Sargenti," with her Italian accent and she did everything to keep me busy. When I would say that I was going downtown, she would say in her Italian accent, "Veddy bad, Sargenti; veddy bad Sargenti" and she would find something important for me to do to keep me safe and out of trouble. I was honored to work with and know such a caring and devoted nun while I was in Japan.

Training the Squad

I was training my squad very hard every day, to prepare them for the hardships ahead. Most of the men were new and very young. I knew that to keep them alive in combat they had to be well-prepared. At first, no one liked me because I demanded a lot during training, but without discipline, they would never survive.

One of the men told me there was an American Indian in the squad who never followed orders and only did what he wanted to do. I wondered how he had gotten this far in the armed forces. I thought to myself, we'll see about that.

A Military Mustang

Having been trained to lead by example, I was always up prior to morning roll call to make sure my cot was tightly made, my boots spit shined, and I was ready for inspection before my squad was even awake. This was my first morning as Sergeant of my squad and I was ready to go when the bugle sounded. When the bugle call came over the speaker, everyone had to get up, make their bed tight enough so a quarter would bounce on it, go to the bathroom, shave, and be ready to fall out.

After the bugle call, I went to the American Indian's bunk and found him still on his cot. I quickly walked over to him, squatted down, and with all my might flipped over the bunk with him in it. He was upside down under his bed and as he came up, he had a mean look on his face. I thought, you don't know who you are messing with.

I came down toward him, no disrespect to the great Indian Chief Cochise, and said, "Alright asshole, Cochise, you get your ass up right now or you will wish you were never born. Do you understand me? Get up, make your bunk, shave, and get out for roll call." He said, "Gee, Sergeant, I was just waiting for someone like you to come along." After that he became one of my best soldiers. Most soldiers automatically followed their superiors. However, some men just need to be shown who is in charge. Once that is settled, they fall in line.

Learning Karate

When I was a paratrooper in Japan, I saw a drunken paratrooper staggering along in front of the shops that had their wares out in front. The paratrooper was across the street from me and with the palm of his hand was pulling all the stuff off the tables and onto the sidewalks, destroying the merchandise displays. He did this at several stores before someone finally called the Japanese police (Kempi).

    The police pulled up in front of the paratrooper. Even with the police there, he continued to knock stuff off the tables. One of the policemen went up to the American and put out both of his hands, palms up going toward him. Not touching him but trying to tell him to stop. All at once the paratrooper, in his drunken stupor, grabbed one of the policeman's hands. The policeman grabbed the paratrooper's hand, made a 180-degree turn and flipped him through a wooden door where he laid unconscious as the Airborne MP's drove up.

    A lot of Japanese people were milling around the area. Being as I saw the whole incident, I walked across the street and said, "Sergeant, I am a witness. This drunken paratrooper pulled all the stuff you see strewn on the ground off the tables, but the policeman never touched him. The officer only put his hands up in an attempt to stop the destruction without touching the paratrooper. When paratrooper grabbed the policeman's hand, in self-defense the officer flipped him through the door. You can see he is out like a light." The officer said to me, "Thank you, Sergeant, for speaking up. We work closely with the Japanese to keep order. It is very helpful when we have a witness that is willing to come forward."

A Military Mustang

The MP called the base to make a report. There were some people in the crowd that understood English and heard me defending the police officer. They saw the Ranger patch on my shoulder. A Ranger to them is like a Samurai. The word was very quickly spread around. "SAMURAI, SAMURAI!"

After everything calmed down, the Japanese policeman came over to me and kept bowing very low in front of me. I felt very ill at ease with him continuing to bow. I said to the MP Sergeant, "Why is he doing that?" The MP said, "You just saved his career by defending his actions and being a witness that he acted appropriately while attempting to stop the destruction. He is a Japanese Karate teacher at the police school and wants you to come to the school and visit." I agreed to go.

They put me in a Japanese police car with the officer and took us to the police station. The man who spoke English came with us to translate. When we got there, I was taken to the big gym where they were practicing Karate. The translator told me that the officer wanted to teach me Karate. I said I would like to learn Karate.

I came for classes at the gym and I learned a lot. The first thing I learned was how to fall to avoid injury when being attacked. Every time the police officer would throw me to the mat, he would say the only two words he knew in English, "So sorry."

My most thrilling experience while at the gym was with an old man, about 90 years old. The old man was the top Karate expert in Japan. They had me wear white gloves and stand in front of him for a ceremony. He laid a 250-year-old Samurai Sword in my hands for a few seconds. Then he lifted it up after giving me a deep bow to the waist. In turn I bowed lower than he did. The police officer teaching me was standing off to the side. After the old man left, he came and gave me a deep bow and, of course I gave him a deeper bow. Giving a deeper bow was a way to show respect and honor in the Japanese culture.

    I am sure the police officer knew when the whole 187th Airborne left overnight for Korea, that we would not see each other again. But I will always remember his kindness in teaching me karate and his only two English words, 'So sorry."

A Military Mustang

Chapter 12 – DAMN 38th PARALLEL

Surrounded

I knew Lieutenant Colonel John Holland personally from serving with him at mass when the Catholic chaplain came on line to conduct a service. I had great respect for him. One time I was out with a patrol of men and we got surrounded. I needed mortar fire to get us back to our own line. When I asked for it, they told me on the phone that they had to get permission from Colonel Holland. I said, "Is he near you? I want to speak to him." When Colonel Holland came on the phone, I said, "This is Sergeant John Arens. We served mass together." He said, "Yes, John, what's your problem?" I explained, "We were surrounded, and I need mortar fire to get my men back inside our line. I wouldn't ask, Colonel, but I'm in a real bind here." He said, "Give me a distance that will not hit you." I told him and they sent one round. It hit about 75 yards ahead of us.

I don't know how many men the mortars killed, but we could look behind us and see the mortars exploding all around us as we moved back to our line. As soon as we moved in behind our troops, the whole line exploded with machine gun fire.

Colonel John Holland was a soldier's soldier. I remember on long marches he would take packs from men who had problems with blistered feet to help them out. One time, I saw him with his own pack and three others on his back. He had pool table legs and was strong as an ox. With his world war battles and three jump stars on his wings, you can believe me, he was another of my heroes. Sadly, after we left the 187th I never saw him again.

Kojedo POW Camp

We arrived on Kojedo Island, South Korea, where Colonel Lee Hak-Ku was the highest-ranking prisoner. He had orders to cause a riot and make it miserable for the Americans. Our General had been released after admitting to mistreatment of the North Korean and Chinese prisoners. A new general, General Boatner, was sent to relieve him. This general spoke Chinese. We were lined up outside the prison on Kojedo Island, off Pusan, South Korea. The night before, we heard screaming from inside the prison camp. Prisoners were running to the fence and climbing up the barbed wire leaving flesh on the wire with every step. We had orders not to shoot the ones going over the first fence. There was about ten feet between the fence they were climbing over and the outer fence that was a kill zone if they were caught. We cut a hole in the fence. Half of us went one way, half the other way. Now we were lined up inside the fence.

Too late, we realized it was a big mistake to give the prisoners a blacksmith shop so they could make tools to work with. They took 50-gallon drums and made homemade bayonets which they attached to long tent poles. It was like medieval armies facing each other only

A Military Mustang

their tent poles were longer than our rifles by about two feet.

We were ordered not to load our rifles at first because they didn't want us shooting the prisoners. Each paratrooper carried about four grenades. Over the loudspeaker, we could hear the commands - drop the tent-pole, take off your cap, throw it on the ground, put your hands on your head. What we saw in front of us was nothing but pure evil. The prisoners who were the hardliners stuck the others in the back with their homemade bayonets. A few hand grenades were thrown by our troopers at the hardliners. The whole operation was stopped when they started surrendering by the hundreds.

It was an all-day affair. We were busy trying to get the prisoners outside the compound into trucks so they could be transported to smaller units where it would be easier to control them.

I was in charge of a group of paratroopers. The paratroopers went into the prisoners' tents. There was a vast amount of junk 12 to 15 feet high. We still had gas masks on after trying to subdue the prisoners without killing them. I was standing next to this high mound of stuff when a paratrooper screamed at me as loud as he could. A prisoner holding a tent pole with a homemade bayonet on the end leaped off the top of the pile. As I turned to look, the prisoner missed me by inches. When he hit the ground, the paratrooper bayoneted him and drove the weapon through his stomach. If my fellow paratrooper had not screamed, the prisoner surely would have killed me. Because we were wearing gas masks, I never saw the face of the paratrooper who saved my life. I hope that someday he will read this book and contact me. He would be about 80 years old now.

We had transferred most of the prisoners outside, and a few of us took one last look before we were going out. I noticed a movement underneath some bedding. Looking closer, I saw a foot and a hand move. Having witnessed what happened to me, the prisoner figured he was a dead man. I knew he did not hurt me. I had seen enough dead bodies that day, so I reached down and touched his hand. I tried to get him to come out. I squeezed his hand three times and he slowly came out. He had a look of amazement on his face. I couldn't speak his language, but I made signs to put his hands on his head so everyone would know he was giving up and no one would shoot him. I pointed in the direction where the trucks were that would take him to the smaller compound, and he walked towards them.

As we continued checking the area, we came into another area where prisoners were tied to tables and tortured by their own people because they thought they would talk. I noticed small freshly dug areas. When we dug into the dirt, we found 50-gallon drums with dead bodies in them. They were all garroted around the neck. Even though I was almost killed a few minutes before, I still felt compassion for the prisoners who were killed by their own men.

After this ordeal was over, the 187th Regimental Combat Team went back up on line and relieved the 27th

A Military Mustang

Wolfhound Division. They were picking up the dead American bodies who were killed shortly before we got there.

As I write this down, I can still see the pictures in mind like it happened yesterday.

Inspecting Fishing Boats

When the 187th Airborne left Kojedo Island they had 100,000 Chinese and Korean prisoners. They were broken into 5,000 compound units. (Compound 76) We were waiting in port. The South Korean Navy had a PT Boat that the United States had given them. I was talking to the captain and told him I was in World War II on tankers and was a Merchant Marine officer in 1949 but was taken off the ship when I was drafted into the Army. He told me that he was going out on a patrol for about three hours. I asked the platoon sergeant if I could go with him. He said we were not leaving the port until the next day, so go for it. He wished he could go himself.

I went aboard with the crew. After we left port, the captain was inspecting fishing boats. However, he would not come alongside the boats because they could fire on them at any time. He went full speed alongside, then made a hard left that enabled them to lean over and look down inside the boat. They were checking out the boats with a 30-degree list. After he did this maneuver with a few fishing boats, he asked me if I wanted to take the wheel and do the maneuver myself. I readily agreed. It was a thrill of a lifetime to steer a PT Boat at high speed. I still remember the thrill to this day.

Deadly Mistake

We went north into combat, my second time. As squad leader, I was responsible for the lives of those under my command. We relieved the 27th Wolfhounds on line. One of the officers over me was Cadet Captain Gerald Schafer. This was our first day on the front lines. Some of the men were new and no doubt were jittery. They would fire their weapon at anything that moved. I always carried small stones in my pocket whenever I was on the front lines. When I checked on my men at night, I would give the password for that night and throw a stone at their position to let them know it was me because the men were even more jittery at night and I didn't want to get shot by friendly fire.

As it started to get dark, Captain Shafer came up to me and said, "As soon as it gets fully dark, I want you to come with me to check on the men to see if they are alert." I told him that it was not a good idea because this was the first night on line and they would shoot at anything or anyone that surprised them. Plus, I can't go with you because I have to take a patrol out in another few hours and I need to get some sleep, first. After napping for an hour or so one of my men woke me up. Shaking me frantically, he told me Captain Shafer had just been shot by one of our men. They were going to take him back to Japan to a hospital. Fortunately, he was shot in the arm, but he could have been killed. He still had to spend about six weeks in a hospital. I went to the man who shot him and told him it was not his fault and that he was an excellent marksman as he hit what he shot at even in the dark. I know he felt a lot better. As a leader, I always did what I could to keep up the morale of the men under my command.

Mine Detectors

Later I was on patrol and heard a guy hollering off to the left of our position. I returned to report it to an officer. He took men with mine detectors to see if they could get to the soldier without stepping on a mine. I used binoculars to watch, but I could see only one dead soldier in the bushes. We kept looking for the one that was alive. I led them to the area where I heard the hollering. The officer went ahead of me into the minefield, then called me over. He obviously had not tripped a mine, so I followed without worrying. The wounded soldier had been hit in the chest and was breathing with a sucking sound from the shrapnel wounds in his chest. The stretcher crew was behind us and they carried him back to a MASH hospital. I don't know what happened after that. I thought that the officer who walked right into the minefield to save a fallen soldier was a brave man.

Chinese Soldiers

Chinese soldiers carried a long slender bag full of rice that went over their shoulder and down the front of their chest. It connected with a part that ran down their back and hooked together. The soldier could open the end in the front and take out a handful of rice to eat whenever he stopped. Sometimes the rice wasn't even cooked when they put a handful in their mouth.

We cooked rice when the wind was blowing toward the enemy lines so the enemy soldiers could smell the rice. We would have a Korean who spoke Chinese tell them to surrender and we would feed them and treat their wounds. First, two men came to find out if we were telling the truth and then more would come. They were a ragged bunch.

When we would get a break from the frontline action, we would guard prisoners in a big camp. The prisoners formed a long line and were carrying rocks to help build a dam. One of the Americans who was not a paratrooper, was kicking every other prisoner in the ass when they would walk by. I got tired of it. So, I moved towards him from the back and put both of my knees behind him and quickly squatted down. He dropped to the ground and I stuck my M1 rifle into his ear and said, "If I see you kick another prisoner, I will blow your head off!" He really believed I might shoot him. I was just trying to get his attention and give him incentive to stop.

An officer asked what was going on. I told him. He answered, "Well he isn't really hurting them." I promptly told him that U.S. Paratroopers would kill the enemy up on line, but once they were prisoners they would not be hurt. It was a matter of decency and ethics. U.S. Paratroopers in the area had no respect for non-Airborne officers in the area after that.

The prisoners saw what happened as they were walking by. An English-speaking prisoner said, "We will work hard for you, Sergeant who stopped the kicking of prisoners." My latrines were dug first and fastest!

A Military Mustang

Chapter 13 – MYSTERIOUS WAYS

Sniper Private Laria

I was staff sergeant and squad leader of the 187th Regimental Front Line. The Chinese soldiers were ahead of us. My sniper was Private Edward J. Laria. He had already killed nine of the enemy with his deadly accuracy. I walked up to his position where he was laying between a bunch of sandbags that provided cover so that the enemy could not see him. I asked him how he was doing, and he said to me, "Sergeant Arens, I am Catholic. Am I going to hell for killing all these men?" I answered, "No, EJ. This is war, they are trying to kill us, you do your job killing them first." Then he asked me if he could go down inside the bunker and get a pack of cigarettes. I said, "Sure, EJ, then get back up here and do what you do best."

EJ ran to get his cigarettes and I started walking away to check on the rest of the squad. When I got about twenty feet away, a mortar round came in right where EJ had been laying and hit right between the sandbags and blew them all to hell. EJ came out of the bunker and said, "Gee, Sergeant, I'm never going to smoke this pack of cigarettes." I answered, "See, EJ, God didn't punish you, he told you to go down and get the pack which saved your life." He thought about that a little bit and said, "You are right." I always wondered if he still had the pack of cigarettes somewhere in his house. EJ survived combat and was credited with 25 enemy-kills.

About that same time, an Army chaplain was walking further down the entrenchments. All at once I saw him start jogging. A mortar round hit behind him where he had been walking. If he hadn't starting jogging, he would have been dead. I walked up to him and said, "Are you a chaplain?" He answered, "Yes." Then I said, looking up at the sky, "Did someone up there tell you to jog?" He answered, "I don't think so." I told him EJ's story about the pack of cigarettes. The chaplain answered, "God works in mysterious ways."

Military Dogs

We had two dogs with handlers assigned to our squad in the 187th Airborne Regimental Combat Team in Korea. They were with us on a mission behind enemy lines. The one handler had a scout dog that was highly trained for smelling and seeing the enemy. The other handler, Jack, had an attack dog. He explained to me that the attack dog was much different than the scout dog. His attack dog would only listen to him. If he should get killed or wounded badly, we should shoot the dog because he would not listen to anyone else. Thank God we performed our mission without casualties and did not have to shoot the attack dog.

One day, I happened to come upon Jack in a rest area. He and a buddy were playing cards. There was a dish of dog food beside his foot. The dog was lying flat on the floor with his nose next to the dish and not touching it. I asked him, "What is going on? When are you going to let him eat?" He answered me, "He is in training!" in an abrupt voice. He continued playing cards with me standing next to him and the dog lying in the same position with his ears perked straight up. He finally said, "Go for it!" in a loud voice and the dog jumped up and started to eat. "By the way, he sleeps next to me all the time. We depend on each other," the handler informed me. So, I learned another lesson in the military, this time about the military dogs we rely on. Dogs are probably one of the greatest unsung heroes. Over the years countless lives have been saved by these selfless canine heroes.

Old Japanese Fisherman

Around October of 1952, after fighting in combat on the front lines against the Chinese and North Koreans with Kim Sung, their fearless leader, we finally arrived in Japan before going to the United States. We had been serving in combat under General Westmoreland.

We were on the pier ready to board a ship back home. Most of the Airborne Troopers had Japanese money (yen) in their pockets. I said, "Why don't we throw all our yen in a hat. Then find some Japanese guy to make happy."

While we were doing that, I spotted an old Japanese man sitting with his grandson on the pier. Most of the troopers going aboard, about 400 of them, really got into the thing of emptying their pockets. Somebody came up with a bag to hold the money since the hat wasn't nearly big enough. I don't know for sure, but there must have been over a thousand dollars. Just before I went aboard, I walked over to the old man, along with six other paratroopers, carrying the bag of money. The old man's eyes almost bugged out of his head.

One of the six guys spoke very good Japanese and told him what was in the bag and that we were giving him the yen because we were leaving Japan for good and going home to the United States. He started bowing to us and we saw the biggest smile with some of his front teeth missing. I felt it was just another mysterious way when things worked out for the best. I knew in my heart we did the right thing when he said he was a poor fisherman and now he could buy a good used fishing boat and make some money. As we walked up the gangway, he and his grandson were still bowing to us, while holding the bag of money. I think to them it was like a bag of gold.

Valuable Allies

In Korea, we had Turkish soldiers and Ethiopian troops fighting alongside us. The enemy was scared to death of Turks because they loved to use the knife at night. The Turks were killing hundreds of North Koreans and Chinese during night. That fact had put the fear of God into them. They were so scared of the Turks, that when they attacked a line and found out it was Turks, they retreated and chose a different area to attack.

A Military Mustang

Chapter 14 – CIVILATION LIFE

Working on the Railroad

I was working on the railroad in Toledo as a fireman on a diesel engine, as steam engines were being phased out. One of the important tests they gave you was an x-ray of your back which I passed with no problems. The x-ray really was not important if you did not have to shovel coal on a steam engine.

The union was strong, making sure only engineers operated the trains, but they allowed a fireman to help because he was useful as a lookout. The engineer sat on the left side running the engine. The fireman sat on the right side. When the train was turning to the right, the fireman on the right, could see all of the cars behind the engine. Of course, the engineer could not see any of the cars from the left side. On a curve to the left, the engineer could see the whole train on his side and the fireman could not see anything.

The engineer was always blowing the whistle whenever we came to a crossroad to give warning to any cars crossing the tracks, especially when the crossing gates were down. The engineer was an old timer and told me he had killed seven people because they drove around the gate right in front of the train. A train does not stop fast; if cars are on the track it will hit them. It is usually deadly if the train is going 40 mph and it hits a car. It is traumatic to helplessly watch pieces fly by your window.

One day, while on the train, I noticed five bags of coal laying on the floor. I asked the engineer, "What's that for?" He replied, "Up ahead, there is an old lady who waits at the tracks for me every few days. Years ago, I used to throw off some of my 'extra' coal for her when I was running a steam engine, but now I am running a diesel engine and we don't use coal. She doesn't know that and thinks I'm still throwing extra coal from the train. So, I buy a few bags of coal and throw them off to her." I thought that was really nice of him.

After that, I worked Hosling diesel engines on a turntable. With the turntable, you could put the engines on different tracks. I was doing a good job for a few days until the boss called me into the office to tell me my X-ray showed I had a bad back. I told them, "No way could I have a bad back."

I went to another doctor and had an X-ray and there was nothing wrong with my back. What happened was the boss was trying to get his son hired on the railroad and switched my good X-ray and put it into his son's file. His son's X-ray with the bad back was put into my file. I raised so much cane, that they fired the boss. At the same time, I quit the railroad because I was not sure what he might do to me to get even. The boss still had friends at the railroad, and it was a 'dangerous' place to work. I did not want any problems.

Construction Work

I was trying to work ashore instead of going back to sea, but the only job I could find was as a laborer. I

A Military Mustang

was getting too old for the hard, physical work of being a bricklayer, electrician, or plumber. On a lot of big construction jobs, even if I was hired for electrical work, I would be the one digging holes, carrying panels, mixing cement, or lining up bricks. If I was working with electricians, I would use sledgehammers with steel bits to break brick walls to run wires. As the new man, I was given the hardest jobs. Really, it was only fair because everyone had their turn when they were new. I was just too old to do that stuff.

While I was there, I had the privilege of working with Big Jake. He was a black man who was 6'6" tall and could swing a sledgehammer all day long and never get tired. Sometimes he would break a big rock in two with just a few swings. When he was a young man, he drove spikes on the railroad. He told me that he was never home. He lived like a nomad in a box car with other railroad track workers. One day they put me to work with him. I was to hold a steel bit while he swung a sledgehammer and hit it. Nobody wanted to hold the bit because he would crush your hand if he missed. After each swing I had to turn the bit a little and then he would swing again. He told me, "John, if you hold the bit very still, I won't hit your hand." Needless to say, I held it motionless.

One day while riding with him I told him that I was an Airborne Ranger and got my jump wings by jumping with the 2nd Ranger Company which was an all-black unit in Korea. He had a lot of respect for me because I jumped with no training. He told everyone how I jumped with the Black Rangers.

One day, we stopped at a bar in a black neighborhood. He told me to go in while he parked the car. When I opened the door and walked in, all I could see were black faces. They were looking at me like, "What the hell are you doing here?" I didn't know what to do. Then in walks Big Jake and he hollered in his booming voice and said, "Don't mess with John, he's my buddy. We work together and he holds the bit while I swing my sledgehammer. You all know I never miss!" One of them said, "You must have a lot of guts because we wouldn't hold that bit." Then they said, "You the man!" They wanted to buy me a drink and I told them, "I don't drink liquor, but I would have a Coke."

Anyway, we continued to work together on other jobs. The women, white and black, used to chase him. But he told me he was married to a beautiful woman and she told him, "If you ever cheat on me, I will catch you, and I will cut it off." He said, "You know what, John, I think she would do it. So, I just ignore all those other women who are interested in me. It ain't worth the risk!"

Interlake Iron

Still trying to avoid the sea, I began working at Interlake Iron as a laborer. My first job was working with the all black group in the hottest place making molds. There was a man with a hammer that knocked the sand off the white mold hanging on a hook that revolved around a circle.

A Military Mustang

The boss was a Hungarian man who ran the place with an iron hand. Being that I was new, and the only white guy, and the boss was not there, I asked one of the black men, "What should I do?" He looked at me and said, "You take the sledgehammer and knock the sand off the molds as they go by you." I asked, "Where is this hammer I'm supposed to use?" The guy said, "Just around the corner." I went around the corner and found a 15-pound sledgehammer and started swinging. I didn't know why they were all smiling at me as I was wearing myself out.

It wasn't long before the boss walked in and all the black guys scattered looking for something to do elsewhere. The boss took one look at me and said, "What the hell are you doing with the 15-pound sledgehammer, you are supposed to use a 5-pound hammer!" Well, I knew I had been had. I told him I took the first one I saw and started working. He had been working at Interlake for 30 years and probably had the same trick pulled on him. For not turning in the guy who told me to use the 15-pound sledgehammer, I was accepted into the group as one of them. I got along great with these guys who worked in such a hot place. They were a lot of fun and I will always remember them as my friends.

One day, a riveter asked me if I would catch rivets. One man heats the rivets and throws them to the rivet catcher (me in this case) who is holding a cone-shaped rivet catcher. I catch it and put it in the hole. The back-upper flattens out the rivet from the other side. The way you learn is they throw cold rivets first to see if you flinch. You always hold the cone away from your face. The rivet thrower is always an old-timer and he knows how to toss the rivet right into the cone. After about 20 cold rivets, he throws a hot rivet, and I mean WHITE hot when you aren't expecting. The reason they were doing this in the shipyard is a lot of the lake freighters sailing in 1962 were made with rivets. Now the freighters are all steel, so they are welded.

Driving a Cab

When I was working at Interlake Iron, the boss asked me if I would mind driving a cab for him. His company was called the Oregon Cab Company. He ran the company from his home. I told him I would, but I needed to carry a gun in the cab. I kept my gun and my money in a cigar box under the seat while I was driving because the lady that had driven the cab before me was shot by a 24-year old. He made her drive him around for about three hours. Then he finally shot and killed her. The cab was number 505. The radio in the cab was the same frequency as the Black & White Cab Company.

I was going to work during the day at the Interlake Iron and driving the cab at night. I was off on Saturday and Sunday. I worked a lot of hours, so I was making good money. One of my important jobs was taking blood from an east side hospital to the west side of the river where there were three hospitals. I would drive full speed straight to the bridge. A police escort would be waiting on the other side of the bridge. I would fall in behind them. I couldn't talk to the police, but my dispatcher

A Military Mustang

could. We would go all the way at 70 miles an hour with their sirens sounding. I had to ride their tails so that cross traffic would not pull in between us. Once when we got to the hospital a police officer said, "You really hung in there close." I told him I used to be a race car driver. The police officer said, "No wonder you could ride my bumper." I was used many times for blood runs.

One day, I was in my house taking a nap. My cab was parked behind my house which was next to the bowling alley. The police saw that it was cab number 505 parked in the dark behind the house. It was known by the police Department that Velma Byers was shot in that cab. Normally, I took the cigar box inside at night, but this time I didn't take the box in the house with me. I went out to the cab, I reached under my seat to check for the box ... no box, no gun! The gun was registered, so I went down to the police station and was told to go to the back room. There was a police officer at a window, and I told him I was the driver of Cab 505. He asked me why was I there? I explained to him about keeping my money in a cigar box under my seat. He pulled my box out of a cage near him and said, "Is this the cigar box that you keep under your seat? The officer saw Cab 505 and thought they were going to find another body. They didn't know you lived next door to the bowling alley. So, they took your box with the money in it." I was not thinking and said, "What about the gun, Sergeant?" He immediately answered in a loud voice, "What gun? You have a gun in that box?" He wouldn't open the box. He kept shoving the closed box at me, saying here is your box with the money in it.

"Are you the driver we escort on the blood runs?" "Yes," I answered. He said, "Just remember, if you shoot a guy who is robbing you, you better make sure he has a gun, or you are in trouble. Now get out of here and out of my sight. Don't forget what I just told you!" Of course, the police officer saw the money and the gun in the box when he inspected the cab, but he did not want to acknowledge that the gun was in the box and he was giving it back to me.

Cyanide

After doing catching rivets for a while, the Boilermaker Union asked if I wanted to take a job about 70 miles from Toledo for three months. The pay was $15.00 an hour. That was big money for a laborer at that time. When I got there, it was a tank farm for a big oil company. I asked, "What do I have to do?" They didn't tell me right off but later someone explained that I would have to work 20 minutes an hour cleaning inside the tank. The next 20 minutes I would be watching the next man inside from the opening. Then 20 minutes off. We would take turns in the tank, working ten hours a day. I asked, "What was in the tank before?" They told me, 'Cyanide!" Now I knew why they were paying $15 on hour.

Blacksmith

I also tried working in a blacksmith shop. The machines we used were called lathes. The lathes were

A Military Mustang

used to turn parts needed for different pieces of equipment. My job was to hold long pieces of metal while the blacksmith made the long rods to be used at the blast furnaces. The blacksmith I worked with was an old Hungarian with 40 years of experience. He turned out a lot of beautiful work. In the 1700 and 1800s the blacksmith was a very important man in many small towns across the county. He put shoes on horses and fixed farm equipment.

During the lunch hour, men would come with metal tins folded over their lunches to heat them so they could have hot food. He would have some fun pretending to burn a hole through the metal, but he really kept an old one out of sight. He would pull it out when they weren't looking, and they would think it was their metal tin. Back in 1962 and early 1963 you could buy dinners in metal containers. After you ate the one from the store you could then use it to bring your left-over supper to work to be heated. All this was before the days of microwaves and plastic containers.

Another tool used by the old Hungarian was a jackhammer run by steam. It had a long lever that would go up and down to bring the hammer down. With his expertise using the jackhammer he could lay a hard-boiled egg on the heavy metal and bring down the moving part of the steam hammer ever so lightly to just crack the eggshell. He let me to try it when I brought my own hard-boiled egg, but I splattered the egg all over. He told me it takes years of experience to do it without smashing the egg. I was proud to work as his assistant.

Cutting Down Trees

After I left Interlake Iron, I was looking for a job when a man asked me to cut trees on his large piece of property. I bought a chain saw from my brother-in-law for $50.00. I told the man that I would do the job using my chain saw to cut down the trees. Then I would burn the debris. It took two months to finish the job. I was dropping trees, making them fall just where I wanted them. I used a sharp axe to cut the branches off and threw everything into the fire. I had fires going all over the acreage. I thought to myself, I am a lumberjack; I know how to fell trees.

A Military Mustang

Chapter 15 – NAVY DIVER

Military Sealift Command

In 1963 I joined the Military Sealift Command. The Navy base in Bayonne, New Jersey, was the Military Sealift Command. I had learned scuba diving as a civilian in 1956. After two years with the Sealift Command, I decided to become a diving officer for Arctic operations.

There are a lot of negatives to diving in the Arctic like water temperature and killer whales. The water in the Arctic is 28 degrees and there are a lot of icebergs, ice masses, and growlers the size of big houses. Growlers are large pieces of ice that have broken off an iceberg. Whatever the size, the ice above the surface is 1/10 the size below the surface. Killer whales eat seals, and divers in black wetsuits look a lot like seals. Need I say more about the killer whales?

I went to the base and reported to a Navy Commander. He looked at my records and said, "You are 39 years old. You will never make it at the Florida Diving School. Navy Seals are the instructors and they will eat you alive." I told the Commander, "I was an Airborne Army Ranger. Don't tell me I can't make it as a Navy Diver. They may throw me out, but I won't quit." He said I could try if I passed the physical at the base and had 20-20 vision. You have to be in perfect shape to be a diver."

Well, I passed the eye test and the last words of the Commander was, "Don't you dare embarrass the instructors by doing more pushups than they do. If you do, you will never get through testing because they will make sure that you don't." I listened to him and made it through the physical testing. I reported to diving school in March of 1965. The young officer looked at my records and said, "You are 39 years old. You won't last the week." I told him exactly what I said to the Commander.

Navy Diving School

My first day in class, a Navy Seal got in my face and said, "I have permission to drown you, sir!" (I knew the young officer told him to get rid of me.) He had pool table legs and a knife tattooed on his right leg and he attempted to drown me. What he didn't know was that I could hold my breath longer than he could, which I did. To make a long story short, I passed the school and he came up to me and said, "If I ever get assigned under you, you won't hold it against me, will you?" I said, "No, because I know that the officer told you to get rid of me. By the way, I can do more pushups than any instructors at the school." I dropped to the floor and did 250 pushups. He said, "No wonder you passed."

First Assignment in the Arctic

I went back to the base and got assigned to Redbud, the ship that was working in the Arctic. It was an adventure that would last almost 11 years. Before it was over, I became a Captain on one of the ships. Every week seemed to bring a new adventure and a new story to tell.

We left the New York port, headed to Argentia, Newfoundland. Our mission was to prepare the port to open. Captain James Hobbs, an old timer in the Arctic, was a rather large man from North Carolina who would break all the rules to get the job done. The civilian divers made $150.00 a day, plus a bonus when they were actually diving. The Navy paid us what amounted to about $7.00 an hour. One of my divers was Eric Khor. He was with me all through my 11 years. When I made Captain, Eric became the Diving Officer. I always admired Eric; he was professional at all times.

Eric Khor played a small part in a World War II movie with Robert Shaw, a great British actor. He stood in a scene where young German soldiers were singing in a cave and stomping their left feet to a German song. Eric was born in Germany in 1940 and was five years old when the war ended. Eric's mother carried him as she ran ahead of Russian soldiers who were ordered to rape every woman they found. During Stalin's time, about 20 million Russians were killed by their own government.

Goat Hill

I completed my first mission in the Arctic water on the ship, Mission Capistrano. Navy trained divers were aboard for diving under ships for inspections. We had handholds welded to the underside of the ship so that we had a way of moving without the ship getting away from us. The Russians would try to pick us up if we came to the surface. A trained diver would be lowered a hundred or more feet below the ship to hunt for Russian Submarines during the cold war.

At one time, the Russians murdered over 4,000 Polish officers and tried to blame it on the Germans who were committing atrocities in the rest of Europe. The town's people, at a place called Goat Hill, told the Americans that it was the Russians that killed the officers. They dug long trenches. To keep the officers calm and docile, they had the officers put on their best clothes, telling them they were going to a new camp.

Instead of a new camp, they took them to nearby Goat Hill. Officers were lined up in front of the trenches already dug. When they saw the pits, they knew they were dead men. It is hard to imagine what they felt when they realized this and saw others who had already been murdered. The order had to come from their infamous leader, Stalin. When the Poles were fighting the Germans in Warsaw, the Russian army stayed on the other side and let the Germans do the dirty work for them, knowing they were going to kill all the Polish officers later.

Putin was a colonel in the KGB. The same outfit that killed their own people years earlier. No wonder the Polish people hate the Russians and Germans who attacked them from the east and the west. Both armies killed hundreds of thousands of innocent people in Poland, especially the Jewish people, in their infamous death camps.

A Military Mustang

John Anthony Arens and Virginia Thomas Arens (circa 1940)

John Anthony Arens, WWI U.S. Navy, 17 years old (circa 1911)

John with his mom while on leave (circa 1949)

Dad with, John, Tom, Barbara, and Delores

John at home with Dad as an Army Ranger (circa 1953)

John home on leave from WWII (circa 1945)

Just before John went into Merchant Marines in 1943

Army Paratrooper (circa 1951)

A Military Mustang

John with a red beard after a long trip to South America in 1947

John as a young officer in the Merchant Marines (circa 1949)

Navigating and shooting the sun, ESSO Bermuda (circa 1953)

Basic training with the 101st Airborne, Kentucky (circa 1950)

Training at the Navy Seals Training School; John is front, left (circa 1965)

Giant Polar Bear looking for some food

John in diving gear (circa 1970)

In Thule, Greenland USNS Redbud (circa 1968)

A Military Mustang

Resting with crew after a nice cold dive

John in the icy water

John and diving team sitting on cake of ice

Arens and Valera

Fifteen-ton anchor from the bottom of the sea was placed at John's door while he was the First Officer in Thule Greenland

Aboard the USNS Blatchford as 3rd Officer in Bombay India (circa 1963)

John with Admiral Graia aboard the USNS Mirfak

Captain John W. Arens (circa 1976)

A Military Mustang

John in Ranger Uniform

John proud to wear the U.S. Navy uniform

John with son-in-law (Lois' son) Matthew Ruch

Ranger Walk of Fame Fort Benning

Arens and Valera

John and Lois Arens - Wedding Day

Captain and Mrs. Arens

Charley Valera and Captain John W. Arens

Chapter 16 – ARCTIC ACTIVITIES

Great Men Who Served with Me
There are men you have met and worked with who stand out in your life and leave their mark on your soul. Looking back, several men stand out in my mind as men who have made an impact on, not only me, but on the lives of many. Chief John Rabbitt is one of those men. Chief Rabbitt was severely wounded in the raid on Grenada to release a group of American students. I met him in 1965. He was an outstanding instructor at the Navy Seals School. It was a tough course, designed to separate the wheat from the chaff. Those who made the grade would graduate from the Underwater Swimmers School in Key West, Florida, prepared to meet the challenges of their chosen field. Chief Rabbitt made sure we were skilled and competent.

Another remarkable man was Rufino Romero. After the Graduation Ceremony in Key West, I reported to the Arctic Operations as the Diving Officer on the USNS Redbud. One of the seamen was Rufino Romero. Rufino dressed me in my diving gear for years in the Arctic Operations. Rufino took care of the bridge, keeping it clean and shining all the brass. He also handled the group of Sea Scouts aboard for a trip to the Arctic. Rufino claimed he had visions. He was standing on the foredeck looking up at the bridge when he saw me dressed in a captain's uniform - at the time I was a Diving

Officer. When I became Captain on the USNS Mirfak, he reminded me of that day years ago when he told me about his vision of me being a captain. We kept in touch after my retirement. He was born in Puerto Rico but brought up in New York City. His sister was married to Red Buttons, the "HEE-HEE, HO-HO" movie and television star.

Another seaman I was proud to have serve with me was Bosun Harvey McCoy from North Carolina. He always seemed to end up on my ship. He was with me on the USNS Rigel when we were in the Red Sea south of the Suez Canal in 1982. He was in charge of the six submersible pumps below decks. I told him if it looked like the ship was going to sink, jump with his men into the water and we would pick them up.

The SS Bahar was saved when McCoy's crew dewatered it after 18 hours of continuous work. I was proud of Bosun McCoy and how he performed the job. I put him up for "Seaman of the Year." When a Spanish seaman won the award for helping during a sea rescue, I called the base and wanted to know how he could be the, "Seaman of the Year," for helping to save a single man in a boat when my guy worked 18 hours to prevent a ship from sinking and saved many men.

The man I was talking to at the base was a black man in charge of giving out the award. I informed him that when I put McCoy in for the award, McCoy sounded too white with his name, but Harvey McCoy was a black seaman. I think there was a gasp on the other end of the

A Military Mustang

telephone. But award or no award, Bosun McCoy is a hero to me and those men who were on the SS Bahar.

Mugged

One afternoon, I headed back to the ship from 54th Street at the MSC base (Military Sealift Command) in Brooklyn, New York. I missed the bus so decided to walk until another bus arrived, then jump on it. While walking down the street, I saw two men walking toward me. This area was known as a bad area (Red Hook Section). When they got abreast of me, they stopped. One of them asked, "You got a dollar, man?" When I said, "No," they each whipped out a switchblade knife.

I noticed a car coming down the street, so I ran between them and out in front of the car, forcing the driver to stop or hit me. My adrenalin had really kicked in. I went around the car and I tapped on the window of the driver's door. There were two black males and one woman in the station wagon. The driver rolled down the window and asked what was going on. I told him very quickly that two guys were going to cut me up with a knife if I didn't get some help fast.

The driver hollered, "Jump in!" I did, right over his lap, past his shoulder, behind his head and down into the back seat headfirst. He quickly rolled up the window. The two men were hitting the side of the car with the back of their knives shouting, "Give us this white MF, we're your brothers," I didn't know if I was jumping to safety or from the frying pan into the fire. I breathed a sigh of relief when one of the men said they weren't going to let them get me.

I came up off the floor into the back seat as they drove down the street until we came to a red light. When I thanked them and got out of the car, I saw a parked police car with two officers sitting in it. I went up to them and told them that I was mugged by two guys with switch blades. The first words out of the mouth of one officer was, "Were they N's?" I said, "Hell no, they were not black, they were Spanish." He said, "Never walk down here alone. Take the bus from the top of the hill down to your ship, unless you got a gun. If you got a gun, shoot both of them and walk away. You will be doing us a favor."

I told the other men on the ship, "Either take the bus or call a cab. Never walk in this neighborhood." By the way, afterward I tried to dive in a car like I did that day, but I couldn't do it again no matter how hard I tried.

The Arctic Circle

When leaving Argentia, the next port was Sondestrom on the Arctic Circle. In arriving at Sondestrom you had to negotiate hairpin turns to get into the fjord. Captain Hobbs, who knew I would be up there for years, would stand behind me and let me take the ship through by directing me how to allow for the turns. Remember, I was 3rd Mate. We would put the buoys out at port and the supply ships and tankers would come to the supply base to pick up supplies for the year.

A Military Mustang

By the third year in the Arctic, I was a 2nd Officer and still with Captain James Hobbs, the last of the experienced captains in the Arctic. I was the navigator, so Captain Hobbs gave me more responsibility. He knew that in 1968 I would have to do some of the captain's work because we would have a new and inexperienced captain. Captain Hobbs had me put all the turn points on the chart during the day. He had also taught me to bring the ship in during fog.

Sure enough, when the next year came, we got a captain with no Arctic experience. In the High Arctic the crew was scared as hell of an inexperienced captain with icebergs all around. This captain was not about to move me up to Chief Officer. He wanted to keep me down, so I told the Command Officer in Bayonne, New Jersey, a Navy Captain, "If the captain on the ship doesn't want me as 1st Officer, I won't be the Diving Officer."

The Navy captain told me that Captain Hobbs said he taught me everything he knew, and I was a fast learner. Remember it is daylight all summer long. So, if I could take a ship in port during the fog, I could definitely take it in during the daylight. In 1968 I was an Executive Officer (XO) on the ship. For some reason, the captain disliked me from the start. Many times, Captain Hobbs used to work me 20 hours a day and let me sleep sitting up in the wheelhouse for 4 hours. This captain was only going to work me 8 hours a day, so nothing was getting done. The base jumped on him with both feet. They told him to work me as many hours as needed to get the work done.

The first problem was that he was afraid of the ice. So, he would not let me come in during foggy weather. When we came into Sondestrom Fjord he almost lost the ship with the zigzag turns, so there was delay, delay, delay. He only lasted one season. If he would have just worked with me, I would have made him look good. I'm not bragging. Captain Hobbs taught me well.

Power Ship

My ship, the Redbud, was called the "Little Flower of the Arctic." It was formerly a Coast Guard ice breaker. She had a 20-ton boom forward of the bridge. When we were in port in Thule, Greenland, the Navy took it over for diving purposes. There was a ship moored behind a breakwater with a lot of boulders and big rocks behind the stern. They built the Air Force base in this area in 1953. It was used to furnish power for the base. Now they had their own power ashore, so they were going to tow this power ship to Guam in the Pacific Ocean and use it there.

A diver was supposed to be coming to Thule, to blow the rocks away from the back of the ship. They were going to pay him $600 dollars to place C4 explosives and blow the wall out without harming the ship. But he backed out at the last minute. So, Captain Hobbs, Commanding Officer of the Redbud said, "Arens, you had experience with explosives in the Rangers in Korea. Do you think you can do the job?" I told him, "Yes." He reminded me they wouldn't pay me because I was Navy diver trained by Navy Seals. I said, "Captain, you tell the

A Military Mustang

Colonel at the base that I will do the job if they will give me a free steak dinner at the Officers Club with ice cream and apple pie."

The base commander said if that's all I wanted it was a deal. So, I checked out the explosives. The only thing the Base Commander was worried about was damage to the ship. I set the charges under the water and when everything was ready, everyone that had been off duty showed up at the port to see the big bang event. Captain Hobbs asked me one more time, "Are you sure you know what you are doing?" I said, "Don't worry, Captain, I won't make a fool of you." I had the wire running back to land a safe distant from the detonator. All I had to do was turn the switch to the right and, KABOOM. All the boulders and rocks blew away from the ship safely. She was towed out the harbor the next day heading to Guam. The Base Commander came to the Officers Club and had a steak dinner with me.

Lobsters

After finishing the Navy Diving School in Key West, Florida, where the instructors were Navy Seals, I was sent to the diving ship called the "USNS Redbud" which had been working in the Arctic for many years. That night after we docked the Captain told me and my team to get pillowcases and go out to the seawall. He told each of us to take a flashlight and get some lobsters. That was my first job as an Arctic Diver with my team.

We proceeded to the jetty which was ice cold. We went under the water and turned on the flashlight between the rocks which made a lobster come out. We grabbed it behind the claws and shoved it into the pillowcase. Between the four of us, we got 70 lobsters and took them back to the ship. The next day, the whole crew ate the lobsters with melted butter. We dove for lobsters many times after that.

I remember the Canadians used lobster pots which were about four feet long with 2-inch slats all around. The top had an opening for the lobster to get in, but the lobster could not get out. A small line ran to the surface which was attached to a floating Clorox bottle to mark where the pots were and to keep them from sinking to the bottom of the ocean. When we were diving, we often found lobster pots at the bottom that had lost their buoy – Clorox bottle. Sometimes we found as many as 30. Each man who owned lobster pots marked their pots so they would know it was theirs. We made a lot of men happy when we brought up their pots and put them on the dock. It was amazing how the men could pick out their own and nobody argued who the pot belonged to if there was a mark.

A Military Mustang

Chapter 17 – NAVY DIVING TEAM

Bergs and Landing Craft

The Navy Diving Team aboard the U.S. Naval ship, Redbud, the little flower of the Arctic, was in the Arctic for many years before I became the Diving Officer in 1965. I continued as Diving Officer until 1975 when I took over the USNS Mirfak. The Redbud was 180 feet long and the Mirfak was 256 feet.

I was manning the ship's Landing Craft Vehicle Personnel (LCVP). This was the same craft that went ashore on D-Day at Normandy Beach. An Admiral had promised that we would get a new reconditioned one before leaving port. I was thankful that he did. A diving officer commanded the LCVP for all diving operations. He would always be the first one in the water with a diver as a partner. This time I manned the LCVP with an engineering officer. To make sure the engine was always in tiptop condition, we worked with the Coast Guard icebreakers diving team alongside our divers. The LCVP job was to not let large bergs the size of a two-story house come together when divers were in the water and block the path to the surface.

I always wore my diving gear when my men were in the water. If the bergs would start closing in on each other as the divers were coming up from working down below on the 4" hose that was used to pump fuel to the beach and the Air Force tanks, it was my job to put the

LCVP between the bergs, preventing them from trapping the divers below. While the divers were down working, the bubbles let you know where they were and when it was time for them to come up. The bubbles also showed when the divers were coming up. I learned to put a 4 x 4 brace in the center when the bergs squeezed the LCVP which is 12 feet wide. I held the icebergs apart with the ramp down. The divers came to the surface and climbed safely into the boat.

Protect Your Buddy

One day, one of the divers was down and I was working with him. He got into trouble below and was coming up too fast. I could see his life vest was getting bigger and bigger. As he was nearing the surface, I took my diving knife out of its sheath and punctured his life jacket because his head was going backward as the life jacket expanded. The air escaped, slowing him down, allowing him to surface safely. He had accidently pulled the side of his life jacket that caused him to rise to the surface.

Before divers got to the surface, they made two 10-minute stops to reduce the chance of getting the bends from returning to the surface too quickly. It was very important to know what to do under any circumstance to protect your buddy when diving. I was with Dennis Langlois who happened to be my cousin. His father had asked me to get him on my diving team. I told him he would have to go down to the Navy Diving School in Key West, Florida. He went and successfully passed the diving school. Then I got him on my diving team as an Arctic diver. Dennis turned out to be a great diver.

Coast Guard Ice Breaker

The USNS Redbud Commanding Officer was Captain Hobbs. At the time, I was in my third year as Diving Officer. The Redbud had lost one of her diesel engines. The Coast Guard Captain of the Southwind asked if I, as 2nd Officer, could pilot their ship into Kulusuk, Greenland, because their captain and other officers had not been there before. The Captain of the Redbud agreed since they were just waiting while the engine was being repaired.

I boarded the Southwind and slept in the XO room as it had two bunks. We went up the east coast of Greenland arriving off the coast of Kulusuk. I told the Chief Quartermaster where to take bearings. There was a lot of ice that year in Kulusuk. Fortunately, the Southwind was a big ice breaker that rode on top of the ice. The weight of the ship would break the ice, so we had no problem getting into port.

Pushup Contest

The Underwater Demolition Team (UDT) goes onto the beaches during a war to clean up charges so they don't explode when boats or men get close to them. In 1967 we were in Goose Bay, Labrador, to discharge some underwater explosives. They were in areas that have humps at the bottom which could cause boats to go aground in the channel. I was the diving officer in charge of a Navy diving team that was working at Air Force base ports. We were setting up the hoses underwater to resupply fuel. Tankers would come in and offload fuel for the winter before everything froze over.

The team had heard that I could do a lot of pushups and one of them challenged me to a pushup contest. All the UDT men and my guys were at the Air Force base. They wanted to do it in front of all the people that were in the club. I told him that since he had challenged me, I got to pick the place where we were going to do it. I said, "We are going to do it in the men's bathroom because I don't want to embarrass you in front of your men and the Air Force personnel." He laughed at me and said, "You are 44 years old and I am in my prime at 24. Who are you kidding, I am not worried?"

But we went into the bathroom anyway and I told him, "You first since you challenged me." He got down and did 247 pushups, stood up and said, 'OK, you do your thing." Knowing he had two witnesses and I had one, I did 420 pushups. When I finished, I told him, "I am not finished yet." I did 100 more pushups every four minutes until I got to 1,100. The 3 UDT divers said, "You are not human. Thanks for doing it here in the bathroom and not out in the club." They were still shaking their heads when we left.

Mighty Mouse

I was First Officer and also the ship's Diving Officer in charge of a diving team on Arctic Operations. The Navy Seals were the best men the Navy had and were

A Military Mustang

proud of the standards set by Roy Bohem who wrote the book called, "The First Seal." I knew him very well. We lived in the same area in Punta Gorda, Florida, and became good friends. There is a post office named after him there. Fred Wright, the Engineering Officer, gave me the name, "Mighty Mouse," because of my small stature and drew a caricature of Mighty Mouse with his fist in the upright position. The ship was called "MIGHTY MOUSE" that day.

Search and Rescue

Sometimes, our mission was heartbreaking. The USNS Redbud was on its third tour of the Arctic Operations. Their first stop was to get ready at the Port of Goose Bay, Labrador, an American Air Force Base. The American Navy Divers were needed to make the port ready to receive fuel for the winter when everything freezes over and cargo ships cannot bring supplies. The divers return from the high northern port in Greenland the latter part of September.

While the ship was in port, a little girl drowned in the Goose River that runs right near the Air Force Base. The river source is Church Hill Falls and it is cold even in the summer months. Seaman Bill Verzi, two divers from Para Rescue Air Force Detachment, and myself were available that day for search and rescue. The young girl's mother was down at the river and when she saw me, she started walking towards me. Her words were, "I know my daughter is dead, but could you please try to recover her body?" Looking at the mother's grief we decided to try, even though conditions were unfavorable for a successful search - there was a swift current, cold water, sunken trees, and visibility near zero.

While we were diving, I was caught in a whirlpool and was sucked down. I struck my head on a rock or stump or something and was stunned. When I surfaced, a Royal Canadian Police Officer in a small motorboat pulled me out of the water. After resting for a while, I went back into the water with Bill Verzi. After six hours of diving the search was giving up. All divers were utterly exhausted. I suggested putting a net across the outlet of Goose River and two days later the little girl's body was found. A proper burial was conducted. The RCMP gave a letter to us gratefully thanking us for the assistance in the operation.

Captain Jack Lawrence

Captain Jack Lawrence was one of my true friends while I was at the Military Sealift Command, "MSC." His career started in World War II after he graduated from the Merchant Marine Academy. He rose in rank very fast. A shortage of officers due to the fast expansion of tankers, liberty ships, and victory ships created many positions for qualified personnel. Jack became the port captain of all the ships at sea while I was still XO of USNS Redbud and USNS Mirfak as Diving Officer.

One day, I made the mistake of saying to him, "I am doing all the work of running the deck department, doing the paperwork, keeping track of all the diving responsibilities for the men in the deck department,

A Military Mustang

working 20-hour days in port, and diving under the ice." He looked at me long and hard and said, "Until you wear the "crown," you don't know what it's like to be the captain of a ship. When you do make captain then you will know what it is like. You will be responsible for every inch of the ship, top to bottom and stem to stern. Your power is absolute. No one can question you while you are at sea. Even the CEO of the largest company in the world doesn't have the power you will have. But, if something goes wrong, you are the one who must make decisions and you better be right, many times the life of your crew will depend on it. Not only that, the crew will be looking at you to see if you show fear. You must be calm without losing your cool in a crisis. You have to think fast, without showing emotion as you give orders. You must also keep the chief engineer informed of what you are doing at all times. The chief engineer is your best and only friend on the ship. Often, he will have many more years at sea than you will as a young captain; but don't forget you are the one with the "crown." If all the power goes off, don't call the engineers, they are trying to find out what is wrong."

    I learned all these things as a Chief Officer. After I became Captain, I kept all this information in mind. When I finally wore the "crown," I understood what Captain Jack meant. I never forgot his advice.

    Captain Lawrence had a 12-year old son named, Branden, who thought I walked on water because he thought being a diver was the coolest thing in the world, especially diving in 28-degree water. He was always telling his friends about our exploits and narrow escapes, particularly when they involved polar bears, killer whales, and leopard seals.

One day when I was visiting Captain Jack's home, Branden asked me to sign my autograph on many pieces of paper, about 50 of them, for his friends at school. Later when I came back from the Arctic, Captain Jack took me aside and told me not to sign any more autographs for Branden. He was selling them for 25 cents apiece. All I could say was, "I think he is going to become a politician someday." Branden was the best PR man I ever had.

A Military Mustang

Chapter 18 – RECOVERY

Rescue Demonstration

The ship's Zodiac boat used by Navy Seals with an outboard motor was in Goose Bay, Labrador. It was making ready at the port to receive fuel for the base and jet aircraft. The base commander was aboard the ship with his wife and son. The diving team was going to demonstrate a fast pickup of a diver out of the water for the base commander and other dignitaries. To demonstrate, the diver in the Zodiac boat was holding a piece of white rope which had been made into an oblong circle. When we did a fast pickup, the diver on the ship would put the diver's hand through the loop and pull him into the boat very quickly. It was used frequently during combat to get men in the water out of harm's way, quickly. The Colonel's son was standing next to his father watching us do the maneuver. When we completed the demonstration, we came back aboard the ship.

When I came up out of the water, I asked the Colonel's son, "How did you like the fast pickup?" His answer was, "You're not so hot. My Dad's a Fighter Pilot!" His father did not know what to do or say. I reached over, put my hand on his shoulder and said, "Son, if you think that much of your Dad, you said the right thing. I am proud of you." I know his dad was proud even though he was embarrassed when his son said it out loud in front of me.

Glacier Girl

While I was in Greenland as a Diving Officer, I heard stories about the B-17 and P-38 Lightnings that landed on the Greenland Ice Cap on July 15, 1942. Authorities thought they must be at least 150 feet under the ice by this time. Not a lot was known about the incident, but we did know they landed before they ran out of fuel and did not crash. The stranded pilots abandoned the plane and were able to make their way to a Coast Guard cutter who picked them up. Between 1977 and 1990 many different teams had tried and failed to recover the downed aircraft.

Two explorers sponsored by the Greenland Expedition found the aircraft in 1988 much deeper in the ice than expected at 268 feet below the ice cap. The explorers carved a long shaft deep into the ice until they finally struck something solid which turned out to be the wing of the B-17. Since the aircraft was badly crushed, they abandoned the recovery effort. Two years later they tried again. It took them almost a month to reach the P-38. The P-38 was found in good condition, so the men went down into the ice cave, took it apart piece by piece, and brought the pieces to the surface.

Pilot, Brad McManus, was able to stand next to the recovered wreckage of the P-38 which had been piloted by his friend, the late Harry Smith. They nicknamed the P-38 the Glacier Girl because the plane was found in the depths of the ice. She actually flew again on October 26, 2002, in front of a crowd of thousands of people. She

A Military Mustang

currently resides at the Lost Squadron Museum in Middlesboro, Kentucky.

Greenland Meteorites

Commander Robert Perry and Matthew Alexander Henson, a black man, discovered the North Pole. They found three large meteorites and brought two of them back to the New York City Museum near Central Park. In 1967, the Danish government asked our captain to retrieve the third meteorite. The United States government turned down the request because it weighed about 32 tons and they were worried about being blamed if it was lost trying to put it aboard our ship, the USNS Redbud. I felt fortunate that I did get to see the meteorite.

Hole in the Hull

I recall the recovery of the USNS John R. Towle after ice punched a hole in the hull. It was one of the funniest operations I remember. It happened on the way to the port at Goose Bay, Labrador for the summer operations. The divers would set buoys and 4-inch hoses to pump fuel from the tankers which would be coming into port. The supply ship, USNS Towle, was named after a Medal of Honor recipient in World War II. It was a victory ship from World War II that the Navy kept for Arctic and Antarctic operations. I was Captain of that ship, in 1979 and 1980.

It was the summer of 1968, the Towle was on its way to Goose Bay, Labrador, and ran into a lot of ice and a small hole was put into the mid-ship area where the plates overlapped. The ship took water in the hold. Lo and behold, the yearly supply of Tampax for the women of the Goose Bay Air Force Base was used to plug up the hole since it expanded when wet. That saved the ship from further damage and allowed temporary repairs while anchored off the coast.

I was flown out in a float plane by an Eskimo man who was taken under the wing of two Canadian float plane pilots. They supported him so he could go to elementary school and finish high school. Then taught him how to fly. The Eskimo pilot told me this story while flying me out to the ship at anchor. We landed next to the ship. I boarded the lifeboat in my diving gear with the Coast Guard divers. After the plate was welded, the Towle proceeded on its mission. My next port was Sondestrom, Greenland, and then up to Thule, Greenland.

In October after completing summer operations in Thule, we went back to Sondestrom on the Arctic Circle. I met the base Air Force Commander told him about the USNS Towle and the situation with the women's Tampax. Well, he told me that the Base Commander was his buddy and he knew him as a brother pilot. He said, "Why don't you let me have one of my sergeants make a plaque with a Tampax in the middle and saying, "Saved by the Women's Tampax of Goose Bay and the date, and I will tell him what we are doing." The wooden plaque was made. When I returned to Goose Bay, I showed the Base Commander the plaque and he was all for it. He told me, "Every week we have a big dance with a band which

A Military Mustang

is coming up in a couple of days. You can present the plaque to my secretary at the dance after you tell everyone what happened to the USNS Towle and how the Tampax saved the ship."

The Base Commander introduced me at the dance, saying I was a Navy Diving Officer at the port. The secretary has no idea that the two of us had cooked up this joke. She was sitting at the table with the commander and his wife. The Commander told her that she should go up to the microphone. I had the big plaque wrapped up in paper, so no one knew what was going to happen.

She came prancing up to the mike in high heeled shoes and I unwrapped the plaque. She saw the Tampax in the middle with the string hanging down, turned around and looked at the Base Commander. I think she was a little embarrassed, but happy, when all of the women and Air Force personnel thought it was the best thing that ever happened at the Goose Bay Air Base. I have no idea whatever happened to the plaque.

Eskimo Huskies

Kulusuk was on the Arctic Circle on the east side of Greenland. The divers were working with 50-foot hoses, 4-inch lines from the water which were laid across a small island where the Eskimos kept their sled dogs. As I've mentioned, the hoses were used to supply the base with fuel. When winter settles in around October, the tankers come in and give them enough fuel to last till the spring when we would return with more fuel. There was always ice in the outer harbor because the water was only 28 degrees.

One time, as we pulled our LCVP (Landing Craft Vehicle Personnel) ashore, I was steering the craft and told one of the seamen to take the line ashore and tie up the craft. He looked at about forty dogs on the island and said, "I'm afraid of the dogs, sir." So, I stepped ashore to tie up the craft and one of the sled dogs came over to me. I put my hand behind his ears. His thick coat of fur felt like a thick rug. As I was rubbing his ears, the large lead dog, "Top Dog in the pack," came up, pushed the other dog away and put his head next to my leg so I could rub his neck.

The dogs loved the attention because the Eskimos did not give them much affection. They considered the dogs were work animals for pulling sleds, not pets. The dogs were treated well. They were fed a lot of fish because that helped them stay strong, but affection was not seen as something that was necessary. The Eskimos would never touch the dogs with a whip, only snap it above their heads. If one dog wasn't pulling, the Eskimo would crack the whip over his head, and he would pick it right up.

Alpine Mountain Climbing Club

Sometimes, it was people we were recovering. During the summer of 1969 the USNS Redbud was assigned to Arctic operations in Kulusuk, Greenland. Navy scuba divers were on the ship to restore underwater fuel hoses and buoys working with Coast Guard divers

A Military Mustang

on the USCG Southwind. Again, working as a team to bring fuel into the bases to last through the winter. I was First Officer and Captain Lyman Couch was Commanding Captain.

The ship was anchored in the inner harbor with a lot of floating ice surrounding the ship. A small boat with a motor came towards the ship carrying four men with an Eskimo (Inuit) running the motor. It did not have much freeboard, the distance measured between the waterline to upper deck level. As it came alongside, the men asked if we would take them aboard. They were from Berne, Switzerland, and belonged to the Swiss Alpine Club that had been mountain climbing in Greenland.

I told them I would speak to the captain of the ship and went to his quarters to explain the situation. The Captain came to the lower deck and we helped them aboard. They had left a ton and a half of equipment on the mountain. Captain Couch said they could sleep in the alleyway with their sleeping bags until we could get them ashore. The Eskimo sailed away happy to get rid of them because of very low freeboard. We also brought all the equipment they had with aboard.

Three months later when we arrived back at the base, they sent two beautiful Swiss cowbells, one for Captain Couch and one for me. They were inscribed with our names, rank, and where the bells were from. The bells had been shipped to Admiral Walter Sheleck, Commander of the base in the New York area for us. The admiral was a decorated World War II commander of a submarine. We reported to his office where he was standing behind his desk with the two Swiss cowbells on display. While we were admiring the bells, the Admiral made a statement, "I wonder how the bells sound." I said, "You are the Admiral, sir. If you say I can ring them, I'll surely ring one for you." He said, "OK," and I rang one. Boy, did it make a beautiful loud sound. I still have my bell.

Pat Thorton from the Heritage Military Museum in Punta Gorda, Florida, located at Fishermen's Village, made a video with him introducing me to tell my story to a couple from Switzerland; Tiffine and Raphel. Pat sent the tape to them so they could check with the Alpine Club in Berne, Switzerland. It would be interesting if they could trace down anyone that was in Greenland in 1969. The men we took aboard were about my age, so they would be retired by now.

Chapter 19 – DANGERS OF ARCTIC DIVING

Diving with Mud

When diving you could see the glaciers ahead except when mud came out from under them. The mud prevented the glaciers from being seen and the work had to be done by feel. As difficult as that was, I was responsible for my men's lives. I was supposed to be their fearless leader. I had to show no fear. Having been an Army Ranger, this was duck soup for me.

The next year was the same, but now I was experienced and wanted to prepare my men for the dangers ahead. On the way north I made all my divers put on their mask so they couldn't see. They had to put bolts in without being able to see while wearing thick rubber gloves. I demonstrated the procedure, explaining that this would prepare them for conditions in the Arctic. Sometimes the water was only 28 degrees. Saltwater freezes at 26 degrees. They didn't like it, but it was necessary. I was as hard on myself as with them.

Quick Removal with Full Diving Gear

Diving was dangerous, but we had some procedures in place to protect the divers. We had a boom hanging over the side of the USNS Redbud and had the same thing on the USNS Mirfak. It was used for getting divers out of the water in a hurry if there were polar bears or Orcas in the area. Lots of times when we were diving through a seal hole, we would put a line through the seal hole down to where we were working. At the bottom of the line coming from the boom was a large sling in the water. All the diver had to do was pass his fins through the sling and sit on a piece of leather about 6 inches wide. He could then be lifted out of the water with full diving gear on and be put down safely on the deck of the ship.

Attracting Sharks

In 1969 I was 2nd Officer (Navigator) aboard the Navy ship USNS Lynch. It was a privilege to serve under Captain Daugherty as he was very much admired. He asked our main base personnel to have me serve as his navigator, which was an honor. Since I only would dive in the Arctic during the summer months, all the rest of the year from October to June, I was on other MSC ships (Military Sealift Command).

The USNS Lynch was scheduled to tow a 350-foot submersible to a specific location. The sub was to be flooded to sink perpendicular with about 25 feet of water above it. The rest would be under the water. Our job was to tow it out to sea and to stand by the operation. It was manned by men who served in submarines. What it really was for was to find Russian submarines. This was during the cold war with Russia when we didn't trust the Russians (Ruskies, as we called them.) You never knew when a Russian sub was lurking somewhere in the ocean. I made it a priority to carry my Arctic scuba tanks and equipment with me on all the ships that I served on.

A Military Mustang

To make a long story short, while towing the submersible, we got our towing line caught in the screw (propeller). No captain of a ship wants that to happen because you are stranded and helpless at sea. Usually a tug has to be called to take care of the problem. When the tug arrives, the divers will go down to clear the line from the propeller. The captain sent a man to my room to wake me up and bring me to the bridge. He told me about the problem. I said, "Captain, I have my diving tanks full of air and I can dive to cut and splice the line so we can proceed with the operation." The only thing I was worried about were the sharks. I asked the captain not to let anyone throw food over the side while I was down under the ship.

Using only a hacksaw that was tied to my right arm, I cut the line where it was wrapped around the propeller shaft. While I was doing that, the 2nd cook on the ship brought out the garbage and dumped it over the stern right next to the captain who was looking over the rail. He couldn't stop him because it happened so fast. As I saw it sinking past me very slowly, I was thinking the sharks would show up any minute because I knew sharks could smell the minutest trace of blood.

When I was almost finished, I met a big blue shark. We came face to face and I snorted into my mouthpiece. I guess he didn't like that, he turned off to my right and disappeared into the depths. He didn't know who I was or what I was, but he left me alone. I finished cutting and removing the line. Now we could maneuver the ship safely away from the submersible. Passing slowly by and throwing a heaving line to the submersible and passing another line from the stern. We were back in business again. We completed the mission and towed it back into port.

When I was back on deck, I found the Captain was mad as hell when he saw the food being dumped. Although, he was happy that there was no reason to call a tugboat when he had divers aboard the vessel.

Looking Like Seals

In the summer of 1968, a Danish schoolteacher asked me if I would bring the diving team to the Eskimo village of Ishlavig. The people there had never seen scuba divers before, and it would be a treat for them. The Eskimo hunters had to be told not to hunt the day we were coming. Wearing the dark diving suits, the men looked like seals. Eskimo hunters shoot seals in the head, so they won't sink. I did not want my men to be mistaken for seals.

At the village, the teacher had all the villagers and school children waiting at the beach. We made sure there were no hunters there. My men and I proceeded to get into the water from the LCVP (Landing Craft Vehicle Personnel) and we swam around with the scuba tanks on our backs. We crawled up on the ice floes like seals and slid back into the water. The people were dumfounded. They didn't know whether we were human or not.

Finally, we proceeded underwater towards the beach and surfaced right in front of the children, spit out our mouthpieces and said, "Hi." When we stood up on

A Military Mustang

the beach to take off our fins, the teacher started to laugh as one of the children said, "It's taking its feet off!" We unzipped our wet suits, about 6 inches, and our body heat caused steam to rise out at chest level. The people were looking, but not sure of what they were seeing.

The teacher explained to everyone who we were and what we were doing here. One of the divers was a black man. The Eskimos had never seen a black person before. In less than five minutes, he had several children around him, each holding one of his fingers and thumbs, and touching his skin. He asked me what he should do. I told him to just smile and enjoy it because he was the star today.

We explained to the teacher how all Arctic divers urinate in the wet suit to keep warm when diving in the 28-degree ice water. School children seem to love to hear little bits of unusual information like this. I've repeated this story many times during my 20 years (1965-1985) of lecturing to school children.

While diving we saw bags of coal on the bottom and decided to bring them to the surface and put them into the LCVP (Landing Craft Vehicle Personnel). This was the same craft that landed on Omaha Beach in World War II. We loaded it up with coal that had been lost and ended on the bottom of the ocean. The Eskimo people were glad to get the extra coal which they could use in the coming winter. We also brought up a few sleds with full loads of coal. I didn't know how they ended up there; I guess they broke through the ice and sank to the bottom.

Cleaning Buoys

One of the most dangerous things that can happen on the deck of a ship is when you are picking up a large weight from the bottom and trying to bring it up on deck. To do this you have to run the wire to a block at an angle. We would pick up a large steel buoy so we could scrape and paint it and put it back into the water. The buoy has anchor chain at the bottom to hold the buoy in place. It is important for the deck sailors not to be in the area where the block could let go. No one should be in that area at any time. If the block let go and anyone was in the area, the wire could snap back and take the legs off any sailor that was caught in the wire snap.

Mirage at Sea

In the Merchant Marines, watch duty was shared on a schedule. The 1st Officer in the Merchant Marine was on the 4 to 8 watch, the 3rd Officer was on the 8 to 12 watch. The 2nd Officer was on the 12 to 4 watch. The 3rd Officer was usually the youngest officer on the ship. He was also in charge of the movies on board.

I was the 2nd Officer on watch 12 to 4 when I noticed something that I was not sure of. It looked like a tidal wave coming at the ship. So, I called the Captain to come up to the bridge. The Captain took one look and said, "Mr. Arens, that is a mirage, but any time you see something like that, call me." That was a crucial lesson. If you are not sure, call the Captain. While a mirage isn't a traditional danger, it is dangerous because men make decisions based on false information. When I became

A Military Mustang

Captain, I said the same thing, "Call me anytime you are in doubt." I would put that in the night orders in the ship's logbook. It is always better to be safe than sorry when men's lives are at stake.

Polar Bears

One of our greatest worries about diving in the Arctic is the gigantic polar bear. The bears use the same holes as the seals and the divers. They would watch the holes to catch the seals when they came up to breathe. The bears would swipe at their heads with their paws to stun them and then pull them up through the hole with their mouth and eat them. To a polar bear, divers look like a big delicious seal. So, we stationed non-divers with rifles to shoot them. We always gave the dead bears to the Eskimos. They were always thankful because bearskins were valuable, and they could sell them at a good price. We felt better knowing that even though we had to kill the bear, at least someone was benefitting. It was not just killing for sport as it was necessary to protect the life of the divers.

Running Out of Air

I was asked to go to all the bases to represent the Navy and take charge at all three ports: Goose Bay, Labrador; Sondestrom and Thule, Greenland. While on duty in September of 1971, shortly before the C.S.S. Hudson arrived in Thule, Greenland, they discovered that their six-foot ram sounding transducer had become bent to the point that it became impossible to retract. This posed a considerable problem as it increased the effective draft by six feet and could curtail the in-shore work of the subsequent program. A further problem was that to remove the damaged transducer would require the services of a very experienced diver.

Luckily at that time, I was a Diving Officer of Military Sealift Command, and was stationed in Thule, Greenland. This was my area of expertise. While I was diving, under the ship and inside the tube with very trying conditions, my diving tanks were somehow caught on something for about 20 minutes. The urgency to get unstuck increased as my air was being expended. It was important to remain calm, so my breathing was not increased, using more air. Finally, I was able to extricate myself and came out safely. Onboard, they had unhooked the ram from inside the ship and it fell to the bottom, about 15 feet under the keel. They threw a nylon line over the rail with a shackle at the end. I tied the line around the ram with half hitches and secured the shackle and the crew brought it up on deck. The crew was aware that I was caught inside the tube for 20 minutes. Of course, they had become very worried until I eventually surfaced and explained just what happened.

It was a very difficult day, but fortunately everything worked at well. I was pleased to receive a letter of appreciation from Captain D. W. Butler, LLD Department of Energy, Mines, and Resources in Canada for what had been accomplished that day in very difficult circumstances.

A Military Mustang

Losing a Diver

I was Commanding Officer of the USNS Mirfak in 1975 on Arctic Operations in Eastern Greenland, an area called Kulusuk which had the same latitude as Iceland. I was the Captain and Eric Khor was the Diving Officer. It was Diver Thomas Yore's first year serving in the Arctic. We had a broken hose and needed a diver to volunteer to go down a line to the hose. Thomas Yore volunteered. It was a simple task to swim out to the hose that was broken in half. Then he was to send a small inflatable buoy up to the surface. Other divers would go down to find the other part of the hose. As I was the Captain I could not dive and had to wait on the ship for details delivered by Walkie Talkie.

I could only assume that Thomas pulled the inflation cords on both sides of his life vest which raised him towards the surface very fast. The bag kept expanding and exploded before it got to the surface. The divers said they saw a large bubble just under the surface when the bag exploded. He was wearing the customary heavy weight belt that helped divers descend. That caused him to descend back down to the bottom once the bag exploded.

A Coast Guard helicopter was flying over the area but could not spot him due to the small broken ice on the surface. They spent a few hours looking for him. I knew that I had to send a flash message to my base regarding the loss of a diver. I informed everyone, including our main base in Washington DC, where our Admiral was in charge.

A team from New York came out to investigate. They first checked to see that all our diving equipment was in good order. I checked with Mr. Khor and he said, "There is nothing wrong with the equipment. You taught me well years ago." The investigating team came aboard along with John Flint, my immediate superior. He was an expert on Arctic Operations and belonged to the Explorers Club. Flint was from Maine and had the accent to prove it. They could not find anything wrong with the diving equipment and the 3-stage compressor used to fill the tanks. Once the investigation was completed, we headed back to New York Harbor, deactivating Arctic ports on the way

When our ship completed its last mission in Goose Bay, Labrador, we finally headed back to New York. When we arrived, I reported to John Flint. I was told that the Admiral wanted to see me. We both knew during the investigation that the body of Thomas Yore was never found for a proper burial. Being that he was in his wet suit he would be preserved for a very long time in the 28-degree water at the bottom of the ocean.

I went into the Admiral's office in New York. During our conversation he stopped and said to me, "Captain Arens, if you had to do it all over again, would you have done it the same way?" I replied, "Under the extreme conditions in Arctic waters with 28-degree water and my experience of running the Diving Team for ten years, I would have done it the same way, Admiral." He said, "Thank you, Captain, and I do appreciate your long service diving in that icy water. Also, your World War II

A Military Mustang

service on tankers carrying high-octane fuel. Plus, your service as an Army Ranger in the 187th Airborne in the Korean War."

I knew then that the Admiral had checked all my records carefully. Now I had to go to Washington DC to meet the Admiral in charge of all the Military Sealift Command and he would also have my records. I flew up from Florida to Washington DC for a Memorial Service for Thomas Yore that was to be held the following day. When I arrived at the head Admiral's office, I saw quite a few men in uniform. I too, wore my white uniform with all my ribbons, medals, Ranger Airborne wings, and Navy Diving Gold Badge for diving.

I noticed the Admiral outside his office looking out at the people waiting to see him. The aide pointed to me, but the Admiral was confused. When he looked directly at me, I raised my hand like, "Here I am." He motioned for me to come to him. I got up and walked towards him. He already knew I was a Merchant Marine Officer with MCS (Military Sealift Command). He said, "I didn't know you would outshine all my Navy personnel. First you have to tell me about your World War II, Korea, and diving in the Arctic experiences. Let's go into my office for coffee and tea." It was hard for him to believe that I went through the Navy Diving School with Navy Seal Instructors at 40 years of age.

The Admiral informed me that I would represent the Navy at the Catholic Church during the Memorial Service for Thomas Yore. He also told me that Tom's father was a lawyer for the government. The service would be held the next day.

We arrived at the church and walked to the front and sat across from the immediate family. After a few minutes the father came over and shook my white gloved hand. We were asked to sit behind the family. I was very honored to be invited. Tom's two brothers were the first speakers. It was such a nice service. I have to admit I cried at the service. Fortunately, they had a box of tissues in the pew. At the end of the service we got to talk to his mother, father, and brothers. Tom died a hero under hazardous conditions as Russian ships harassed us every chance they could during the cold war.

I went back to my home in Florida and got ready to go up to the Arctic again the following year, continuing the same work with the divers. We were a good team for many years.

Chapter 20 – EVERYDAY OPERATIONS

Submarine Diving

When I was a Navy Officer Diver, we only could dive in the Arctic from June to the end of September due to the freezing water later in the year. I had a chance to inspect submarines that were sitting in port and check to see if they had any divers aboard. If they didn't, I'd offer to inspect the hull and propeller to see if anything was caught in the screw. When I was swimming around the hull, I would tap using my knife, DOT-DIT-DIT-DOT-DOT as a kind of signature. After the inspection I got out of my wet suit, put on my uniform, and they'd serve me a wonderful meal on board their boat. While eating aboard I would tap the same signal with my knife and then they knew I was the diver outside. The food was always great on a submarine.

Liaison Officer

In 1972 I was the Liaison Officer in Thule for the Navy. The Air Force was in charge of the base, but the Navy had the responsibility to take charge of the pier. The Coast Guard Cutter Southwind was assigned to break ice in Thule for any ship coming into the pier and to help offload, if needed. We (my team of Navy divers) were working with the Coast Guard divers inspecting under water, we found deterioration of the pier. The Coast Guard cameraman who was under water with me helped to assess the damage by using a yardstick painted white with large black numbers that would determine how badly the concrete had been lost through the years. Then we would make our report.

An Air Force Two-Star General arrived for a visit and inspection. There was a meeting at which time I was going to show him the pictures of the pier damage. Just before the meeting I was introduced to the General. When he saw my ribbons with the Combat Infantry Badge, he was confused why I would have that on top of my ribbons. I also was wearing a Navy Divers Badge that was received in 1965 when I went through the Navy Seal Training School. He asked me what combat outfit I served with in Korea. I proudly said, "I served in combat with General Westmoreland in the 187th Airborne Regimental Combat Team." He came back with, "I was assigned to Westy (as he was called by higher officers) as his personal pilot.

After that I could do no wrong. I gave a presentation showing the degree of the damage to the pier with pictures to show the details. When I was through, he said, "I will contact your Admiral and between us we will get the pier fixed, whatever it takes. So, you have been up here every summer since 1965? That is amazing. How cold is the water you dive in up here?" asked the General. I answered, "28 degrees." He said half to himself, "I'll be damned!"

After the Air Force General left, I was invited by the Air Force Chaplin Captain Bobby Black, to go up to Kanak north of Thule. That's where they moved the

A Military Mustang

Eskimo village when they built the Air Force Base in Thule, Greenland. This was done to keep the Air Force personal from getting involved with the young Eskimo girls. This was a much better idea than trying to monitor activities between the young men and the girls.

Before we boarded the plane, Chaplain Black told me that the locals would come to the plane with their children. When we arrived, I would have to kneel down in front of the children and make faces. The more faces you make the higher your esteem will be in their eyes. When we landed, sure enough the children met the plane and Chaplain Black and I knelt down and proceeded to have a face-making contest with the children. I performed the most faces and won the contest. The chaplain said, "John, you are really in with these Inuits." When Chaplain Black told them that I was the Navy Diving Officer since 1965 and had to dive in 28-degree water, I was an even bigger hero to them. It was great fun and a pleasant diversion from the everyday dangers of diving in the Arctic.

The Little Things

Traveling, we met a lot of interesting people from various countries. In Portugal we met the dory fishermen who worked alone in the icy waters. They were fishing for codfish with long lines that went down into the Arctic Ocean. It was a continuous line that had fishhooks attached. They would take a fish off, bait the same hook and feed the line back down to catch more fish. When their boat was loaded, they would row back to the mother ship that was painted white. The mother ship was a sailing ship with an engine.

We found we could help people in small ways. For example, our ship would save empty jelly jars. We also collected them from other ships when we were in port. People back at our base in Brooklyn, New York and the Bayonne Naval Base would also save the jars for us. Whenever we would see a lonely dory man, as we called them, we would put a pack of cigarettes and a pack of matches inside the jelly jar and then seal the top. Being that his fishing line went up and down on a continuous line, we could go pretty close to him and drop a jar without getting tangled in his line. He only needed to row a short way to pick up the jar. We would put our hand to our mouth, going out and back, mimicking the motion of smoking a cigarette. He got the message and waved back to us as he rowed over to pick up the cigarettes.

After doing this for quite a few years, I met the captain of the mother ship. I told him who I was, and which ships were involved in this project. He had quite a few dory men on board who came to the bridge to meet us. They said, "We always wondered who was giving us the cigarettes." I told them I was the Diving Officer on the ship and that we dove in 28-degree water and worked on the pipelines for the Air Force. They said, "We thought we had it hard but not after your story." Life for the dory men was difficult spending their days alone. Some dory men were lost due to the fog and killer whales that dumped the boats. We were glad to be able to help in some small way.

A Military Mustang

Gurkha Soldiers

The USNS Blatchford C-4 troop ship was carrying Indian troops back to India. The Indian General saw my ribbons and had me sit at his table with him for meals once he found out that he and I were in the same area during the Korean War. His charming wife stuck with him when the Chinese Communist took over the country in 1946. The Chinese came into the Korean War during the winter months. That was brutal for the Army and Marines because they did not have appropriate winter clothing, but the Chinese did. When soldiers killed the Chinese, they would take their winter uniforms to keep warm.

The Chinese Communists were backing Kim Il Sung who was the grandfather to the existing leader, Kim Jong-un whose father was Kim Jong-il. Dennis Rodman, a famous basketball player, visited the leader and embarrassed himself by having pictures taken while dressed up like a bride and sitting next to the leader and calling themselves best buddies.

Kim Jong-un claimed he was going to attack South Korea with his million and a half-man Army even though he couldn't feed his own people, who were as poor as church mice. Our country and others were sending food so his people would not starve to death. Yet, you see thousands of soldiers in his Army looking really healthy when they would march past all the generals and the leader. Because they were receiving so much aide, he was able to use more money to support the army.

The Gurkha soldiers onboard the ship from the Himalayan Mountains did not look like Indians. They look more oriental. Even though they are small, they don't have the word, "retreat," in their language. They carry their rifle and a kukri. A kukri is a Nepalese knife, that has the weight in the front part of the blade which is thicker than the handle. It is deadly and can very easily lop off a head with one stroke.

When we arrived at the port in India, I noticed a general in the Indian Army salute this Gurkha sergeant. I asked the British officer, "Why did the general salute a sergeant?" He answered, "In World War II, the Gurkha sergeant beheaded 25 Japanese and always laid the head on his right shoulder so the Japanese would know it was the same man doing it. During World War II, he was awarded the highest honor bestowed by Britain on its military by George VI, Queen Elizabeth's father. I was honored to meet the Gurkha sergeant on the dock and proceeded to salute him and shake his hand. He had a great smile. All the Gurkha soldiers admired him greatly, as did I.

Whitey Stakokis

Whitey Stakokis was one of my top divers. He was also a ski instructor at Mount Stow, Vermont. His favorite beer was *Heineken* and it was the only beer he would drink. We were in port at Thule, Greenland one day. He left our ship at the pier and went to the club. He ate dinner there and since we were not diving that day, I told him to have a good time, meaning he could have a

A Military Mustang

beer if he wanted one. He came back to the ship about an hour later and told me the manager kicked him out of the club. I said, "Well, I'm going up that way and I will stop at the club and find out what is happening."

I drove up to the club and I saw the manager. I asked him, "What is the deal? Did you kick my diver out of the club?" He said, "Yes, I did. He ate three full steak dinners with all the trimmings and was ordering a fourth one. So, I kicked him out of the club and told him to go back to the ship because someone else might like to eat a steak dinner too."

Later when I came back to the ship, I ran into him and said, "You didn't tell me you already ate three steak dinners and had ordered a fourth one." Whitey said, "Gee, Mr. Arens, I didn't know it was a problem as I was paying for all them dinners." I'll tell you, Whitey could eat, eat, eat and never put on a pound.

Navy Pier

After spending many years as the Diving Officer in charge of an Arctic Navy Dive Team with the USNS Redbud and Captain of the USNS Mirfak, I was asked in 1973 to be the Liaison Officer in the Arctic Operations at Thule, Greenland. I'd be in charge of the Navy pier and all ships that docked at that pier. My quarters were very nice, and I had a Navy pickup truck at my disposal. Next to the flagpole, in front of my office at the base there was a 15-ton anchor from an aircraft carrier.

One of the heavy-lift ships pulled up two lost anchors from where the ships used to anchor near the pier. During a heavy storm the ship moved dragging the anchors through the pipelines. After that, the tankers had to "med moor" which meant dropping two anchors and go astern close to the pier with the stern backed up to the pier. Perpendicular angle parking basically.

While assigned there, I had my diving equipment with me and filled my air tanks from the base firehouse. One of the Danish machinists made me a one-of-a-kind adapter to go from my tanks to the large tanks in the firehouse. The adapter allowed me to get an unlimited amount of air anytime I needed to. I also had a compressor on my ship. I had two anchors on the pier. I gave one to the Danish Liaison Officer and one to the Colonel in the Air Force for the big building that was connected to large antennas facing north that were used to guard against Russia firing a missile towards the base. This was called the Dew Line. We had anchor chain, so we put the chain right up to the building.

Standing in Christopher Columbus' Footsteps

My ship, the USNS Wyman, was in Barcelona, Spain. I was the First Officer. The Commanding Officer was Captain Broom. We were staying in a first-class hotel until one of the Spanish guests said they were moving to a much nicer hotel not too far away. Many of the Spanish people stayed there and it was much cheaper. After staying one day at the first hotel, I spent the rest of the time in that very beautiful old hotel. While eating lunch the manager came over to me and said, "After you

A Military Mustang

finish eating, I want to show you something. It is nearby, within walking distance."

After walking only about two blocks we turned into a place that had a beautiful building in the distance. We walked up to this building that is now a museum. A curator opened one of the large doors and came out. The hotel manager told me that the curator was his friend and he was going to open the museum just for me and give me a private tour.

We started up the steps when the man said, "Stop right there! Put your left foot there and your right here." It seemed like a weird request as I couldn't figure out why I should do it. I was confused. The curator explained that I was standing in the same place that Christopher Columbus had stood when he met Ferdinand and Isabella, King and Queen of Spain in 1493. I was thrilled to be standing in the same footsteps as this famous man. I remembered seeing a painting of Columbus meeting the King and Queen of Spain when I was a young boy. Then I looked up and saw the small window that was in the painting, where a young page boy was looking out the window. That was the painting that was in our history books. Knowing the history made it so much more impressive to be standing there.

The curator took me inside. After a tour of the museum, we went back out the large doors and started walking back towards the entrance of the hotel. The manager turned on a switch that lit up everything about 25 feet down. We were under the palace and saw all the old Roman ruins. The castle was built over the top of them. It was really interesting to look at all the Roman statues and the many beautiful Roman works.

Chapter 21 – TRYING TIMES

Trouble on the Lake

I was living in Lakeworth, Florida, with my father-in-law, Earl Keeling. He loved to fish, and I had driven him to Okeechobee Locks. The gates were opened about five feet allowing rushing water to come into the river from the lake. As I watched my father-in-law fish for a couple of hours, I noticed a large black lady fishing about 30 feet from us. All of a sudden, the woman stood up and turned away from the wooden wall she was fishing from. As she did, she lost her balance and fell backward into the swirling water below. The rushing water formed a whirlpool and it was carrying her towards the middle of the lake.

I took off my shoes, sunglasses, and hat as jumped into the lake because I could see that she didn't know how to swim. Her head was bobbing in and out of the water. As she was going under each time with her mouth wide open, she had to be swallowing water. As I was getting closer, I approached her from the back as lifeguards are always taught to do. If you approach a drowning person from the front, the person will panic and grab you. The person will try to climb on top of your body to get out of the water. Of course, you both would be in trouble then and could drown. I hollered in her ear, "Don't fight me, I will save you!" Once she felt safe with me and quit flailing, I began swimming toward the bank.

Three black men who had been fishing nearby, rushed over with their fishing poles. When I got close to the bank the men extended the fishing poles down into the water for her grab onto. She was extremely tired but fortunately she was able to hold on. When I got her next to the wooden wall, they got hold of her arms and pulled her out of the water. By then, I was very tired because I was carrying someone while swimming against the current. So, I needed help to get out of the water as well.

My father-in-law was oblivious to what was going on and just kept right on fishing. When I got out of the water, I went over to the lady who was lying on the ground. A lot of people had shown up and were on the shore watching what was taking place in the water. The first words out of her mouth were, "Oh, you are my lucky day." I found out she lived alone and fished every day to eat. She held my hand for a long time and didn't want to let go. She told me that she didn't know how to swim and would have surely drowned if I hadn't jumped into the swirling water to save her.

When I talked to my father-in-law, he said to me, "There sure was a lot of commotion over there". Standing there in my wet clothes, I told him I just saved that a lady from drowning. As we drove home, I felt pretty good knowing I saved someone from drowning.

Flying with Colonel Balchen

Colonel Bernt Balchen, now 71 years old, arrived at the base for a visit as a civilian. He was the Colonel in Command of Sondestrom Air Base during World War II.

A Military Mustang

He was a famous bush pilot in the Arctic and had nerves of steel. In 1929, Admiral Richard Byrd chose Balchen to be the pilot who would fly him to the South Pole, because he was qualified to fly using instruments. (There were a lot of World War I pilots who flew biplanes, but they were not trained to use instruments.)

Air Force personnel asked me to escort Balchen around in my truck, taking him anywhere he wanted to go. While we rode together, I had a chance to question him about flying over the South Pole. The tri-motor plane he flew in is now sitting in all its glory at the Henry Ford Museum in Dearborn, Michigan.

Balchen explained to me, the great pilot, Floyd Bennett had flown Byrd over the North Pole but had health problems, so he had to look for another pilot, "That's how I made my famous flight."

We had taken off in the lower part of Antarctica at Byrd's Station. I was the pilot, Bernt Balchen explained. Everything was going well until we came to a glacier. Remember the South Pole is landmass and we took off from sea level. The landmass had glaciers over higher landmass. As we are approached the front of the glacier, I told Byrd to throw out all the equipment. As we continued, I estimated we were about 20 feet from the top of the glacier, way too close to continue safely. So, I told Byrd to throw out the survival safety equipment, too. Concerned, Byrd replied, "But we won't have supplies to survive if we crash." I said, "If we don't lighten the plane, they will find our bodies in the plane at the base of the glacier years from now with the safety equipment still on board." After getting rid of the extra weight, we hit the updraft that I had anticipated and over the top we went. That was a very harrowing flight to the South Pole. Bernt said he flew around the South Pole area, then headed back over the glacier. It was all downhill flying back to get back where we took off.

When Balchen was colonel at Sondestrom, an aircraft landed on the ice cap east of the base with engine problems. The Colonel got up from his desk, jumped in a plane and flew out on the Greenland ice cap. After the injured aircraft landed, he kept his aircraft circling on the snow slowly. He couldn't stop because the skis would stick to the ice. The crew from the downed aircraft jumped in the door of the DC3 as it circled and then he flew back to the base. The Colonel went back to work in his office just like nothing happened. He is what I would call a man among men.

I was proud to have met Colonel Bernt Balchen and to have accompanied him. I showed him the empty silos where missiles had been. (The missiles were removed the year before I came to the area.) I had my camera with me. At the end of our time together, I asked the other person with us if he would take my picture with the Colonel. I cherished that picture to this day as it is not every day that you have your picture taken with someone famous.

USNS Mirfak Arctic Operations

Upon arriving at the entrance to Kulusuk Harbor we saw a red Norwegian ship caught in the ice. The men were playing ball on the ice next to the ship. I called to the vessel on the ship-to-ship radio and asked the captain if he needed any help. The captain said yes, that they had been trapped in the ice for a few days. I told him we had an icebreaker bow. The Mirfak proceeded to break the ice across the bow of the Norwegian ship as close as safely possible without hitting his ship. The ice broke and we led the ship into port.

After arriving both of us dropped anchor and continued to talk on the ship-to-ship radio. I let them know we had a movie about their ship with an American actor, Alan Ladd, the 1st Officer on their ship. They sent a boat over to pick up the movie. They played the movie day and night until everyone had a chance to see it. As the ship departed, the Captain said, "Thank you for everything you have done for us and we will never forget you and your ship."

The Russian Fleet

When we were in the Arctic in 1976 during the summer months, we happened to run into the Russian fishing fleet. Everyone with naval sense knew the fleet had an admiral in charge and a political officer. The 3rd Officer on our ship, the USNS Mirfak, was Polish, but he could speak excellent Russian. He was a long-time sailor with much experience. However, he hated the Russians even more than the Germans because it was the Russians that split up his country. Every day he had to listen to the Great Russian System.

He shared some of the history with me. Putin was a colonel in the KGB of Stalin, who ran everything. Stalin killed 20,000,000 of his own people through starvation and the Gulags. Putin came from the group at the Gulags. One of the Doolittle Raiders landed in Russia with a B-25 which the Russians confiscated, and it was never seen again. When the Russian fishing fleet was around us, my 3rd Officer pretended he was their admiral in charge and gave them orders in Russian. The Russian admiral would holler over the radio, "I'm the admiral." Our 3rd officer would holler over the radio, "I'm the admiral. Don't believe him!" The crew claimed I was playing soccer ball trying to keep them from hitting our ship. They never did touch the USNS Mirfak.

A Military Mustang

Chapter 22 – MANY DUTIES

Captain Manual "Manny" Vieria
After making Captain, I was now being asked by the base to drop back to First Officer. My friend, Captain Manual Viera had requested me as his First Officer. I was his First Officer when he was Captain of the USNS Rigel. I admired him as a man and a friend, so I was honored to be asked to serve under him. Captain Vieria's family immigrated to America sometime in the 1800's. He could speak three languages fluently - Portuguese, English, and French. When he ate at the Captain's Table and I sat next to him as his first officer, he taught me a lot about being a good captain. Not to mention his eating habits, European style, were something to see; so fastidious and neat. He loved taking care of the flowers around his house. He passed away in Florida after a long illness in 2011. I lost a great Captain and my best friend.

Helicopter Operations School
As First Officer, I was assigned to the USNS Rigel for underway replenishment at sea with Navy ships and was directly under Captain Manuel Viera. To see that all deck personnel were trained properly we had a retired Navy Chief Bosun Mate who really ran the show while in port. I reported to the Norfolk, Virginia, Helicopter Training School to learn how to propel down a rope or jump from a helo hovering about 30 feet above ground.

Pilots would not go lower because they did not want saltwater in their engines. I also learned how to bring a helo down on the deck of a rolling ship. I had already jumped out of the Coast Guard helo in the Arctic with floating ice all around, so that was easy.

Commendations for Navy Divers

The Navy Divers that served with me enduring extreme conditions diving under the Arctic ice deserve commendations. I commanded a team when the tanker, SS Potomac had a large 15-foot hole in the bow caused by ice. Being the Diving Officer, a plan was implemented to keep the bow in the down position and pump out the after-tanks first. When more tanks came forward, they were pumped out and the bow stayed down so that no oil could get out. Later a submersible pump was used to pump out the tank in the bow. When operations were finished, it was reported to the Danish Government at the MSC Navy Base in New York that all was clear.

Some of the divers that worked with me were:
Erich Koher, Third Officer
Whitey Stakokis, Able Bodied Seaman
Dennis Langlois, Able Bodied Seaman

I trusted my life with these divers who were trained by Navy Seals. It takes a man with special qualities to dive under the Arctic ice. Every year we worked with Coast Guard Divers as welders using a burning torch to complete many jobs.

A Military Mustang

The New GPS (Global Positioning System)

I used one of the first GPS (Global Positioning System) but still used the regular compass and navigation system. Anyone that sailed in the Arctic knew that the closer you got to the Arctic your magnetic compass was basically useless. It was the Sperry Gyro Scope that still worked. Little did anyone know that the GPS would become so common it would be a standard app on personal cell phones in the future.

Argentina Steak

I was the Commanding Officer of the USNS Mirfak when it sailed the Eastern Coast of South America to Bahia Blanca, Argentina. The ship was carrying a full load of ammo which would be dropped off at the Navy Base. The Navy always treated us very well and they had a band playing on the dock when the ship was being secured to the pier. I was called on the ship to shore radio and asked to inspect the troops. Apparently, they were informed I was an Army Ranger and Arctic Diver trained by Navy Seals.

Argentina was known for the best beef. The gauchos herding the cattle on the pampas of South America, became experts with a *bolas* and a knife. The Navy Commander sent the best steaks to the ship. The whole crew really enjoyed them and there was even enough for seconds.

After the ammo was unloaded, we left the port and went south to the tip of South America. We proceeded to go through the straits with a pilot who was a retired Admiral. He said it took Portuguese explorer Ferdinand Magellan thirty-eight days to navigate the straits since he was the first European to do so. Magellan took many wrong turns which cost him many days. He started with five ships with him. One deserted and headed back and one was wrecked. The remaining three ships reached the Pacific on November 28, 1520. The strait was named the Straits of Magellan.

Saluting the Esmeralda

The pilot stayed with us and we took the inner passage quite a way up the western side of South America. It seemed like a sightseeing tour as we passed many Indian villages. We then headed for Valparaiso, Chile. This was a beautiful town and we got to see their Army Base which is equal to the U.S. Army's West Point and the U.S. Annapolis Navy Base. We were shown many artifacts displayed in their Naval Museum from the Esmeralda, a Chilean Navy ship which had battled with a Peruvian ship. As the story goes, the Esmeralda was sinking, and Captain Prat gave his final order to jump onto the Huascar with swords and knives crying, "Let's board, lads!" The wrecked Esmeralda was made of wood and the Peruvian ship was made of iron.

Coming into port I asked where the buoy was located that was placed over the wreckage. The pilot gave the location and I told him that we were going to stop the ship over the Esmeralda and drop the American flag as a salute to their sunken ship. Afterwards, the pilot called

A Military Mustang

the base and told the commanding officer what we had done. When we arrived in port there was a big reception waiting for us. A military reception, complete with a band. The men were all standing at attention on the pier. The pilot also told the colonel in charge that, "Captain Arens was an Army Ranger, 3rd Company, and 187th Airborne, and a Navy diver in the Arctic." They were amazed that the Arctic Diving Team was diving in 28-degree water. I was asked to inspect the Army troops. It should be noted that the men were not walking during the unloading of the ship; they were jogging the whole time. They put on quite a show. I commended the men for their dedication.

Smart Ship Award

I became Captain on July 7, 1975, by the Admiral of Military Sealift Command. I would have never thought I would be Captain on a Navy ship. It was ironic because I was not able to join the Navy when I was younger because I had already been drafted by the U.S. Army.

I was commanding the USNS Mirfak at the time of the presentation. It takes a whole crew to make a great ship and the crew worked very hard to make the ship First Class. I was proud of my men and their efforts. The Smart Ship Award was presented to the USNS Mirfak before it headed back to Arctic Operations. The prestigious Smart Ship Award goes to the top ship in Military Sealift Command. I was honored that Captain Seamon was sitting next to me when the ship was presented the award.

Captain Seamon had just retired shortly before the ceremony and wanted to be there when the award was presented to the USNS Mirfak. The MSC ships are Navy ships manned by Merchant Marine officers.

Chapter 23 – PROBLEMS

Misunderstanding

Our ship, the USNS Mirfak, named after a prominent star, was sitting in the New York port. One of my seamen who was born in Puerto Rico, had a very strong Spanish accent. He came to my room and was explaining to me that the base was going to let him go because he lied on his application by not knowing what the word, felony, meant. I asked him, "What crime did you commit?"

He started explaining to me that his sister worked in a neighborhood store, back home in Puerto Rico. A man came into the store to rob it with a box cutter in his hand. When she resisted, he cut her cheek deep enough to also cut her tongue which prevented her from talking plainly. I said to him, "What did you do?" He said, "I hunted him down because my sister knew who he was, and I killed him." That was the reason he had to get off the ship.

I went over to the base to the head civilian who was the person in charge of hiring. When I arrived at his office, the secretary passed me through. When I entered his inner office, he had two assistants with him. As I was a captain, he welcomed me and said, "Glad to see you. What can we do for you, Captain?"

I started by saying, "If your sister worked in a small neighborhood store and a guy walked in and tried to rob the store, cut your sister's face with a box cutter deep enough to cut into her tongue, what would you do?" He answered, "I would kill that SOB!" I replied, "That's exactly what Gonzales did, sir, the guy you are going to fire because he didn't know what the word felony, meant in English. His two assistants started laughing and he hollered at me, "You sandbagged me, Captain!" And I left.

We made a short trip and when we came back into port, I was sitting at my desk with the door open. Gonzales came sliding into the room on his knees and grabbed my hand. He started kissing it, "You saved my job, Captain Arens." I was hollering, "Get up, get up!" I was afraid someone would be walking by and see him kissing my hand. I told him I was very happy for him. Remember, the Lord works in mysterious ways!

Ingenious Docking for Earthquakes

In Valparaiso we were tied up to the dock. They used wire, but in the middle of each line, they had about 20 feet of nylon line as a safety feature that provided a stretching factor. Sure enough, they had an undersea earthquake about 20 miles off the coast while we were there. The harbor felt it as the ships were moving at the dock. That nylon line attached to the wire saved our ship from tearing away from the pier. If it would have been just wire, the ship would have broken loose in the rough water. Valparaiso has more earthquakes than any port I had ever visited. So, they developed that method for

A Military Mustang

securing ships and it has been working for years. The crew and I enjoyed the whole trip.

Respecting Customs and Traditions

I was the Commanding Officer of the USNS Wilkes, reporting to the ship August 1, 1978. Pakistan is a Muslim country that was part of India after World War II and for the first time, it was a country of its own. When I reported to the ship to relieve the other captain, I immediately went through the ship's records. Reading the records, I realized we were missing some very important parts for the ship and I found they were tied up in customs. I put on my white uniform and asked our liaison to make an appointment for me to visit the Admiral of the Pakistani Navy. I was quickly notified that I could meet him that morning. He would be glad to meet the captain from an American Navy ship.

He sent his personal car with his personal driver who spoke English to pick me up. When I arrived at his office, I was greeted with a lot of fanfare from his staff standing off to his left. I wore all my ribbons on my uniform - Army Ranger Badge, Navy Parachute Wings, World War II ribbons, 187th RCT, and Navy Divers Badge because I had learned over the years people are more likely to do what you want if they think you are important. Most notably on top was my Combat Infantry Badge that you could only receive if you were on the front lines facing the enemy in hand-to-hand combat. Before we got down to business, I had to explain what was on my chest and how many battles I was in. He was really quite impressed. They had not met a captain of US NAVY ships with that kind of decorations before.

When all of his staff left and it was just the Admiral and me, he asked if I would like a cup of tea, accompanied with sweet rolls. Then the Admiral asked, "Would you allow me to pour the tea?" When I picked up the sweet roll, he noticed that I was sitting on my left hand. He said, "Why are you doing that? I answered, "I am an American. I do not want to offend you or make the mistake of using my left hand to eat with." Now he was really impressed. He said, "You are the first American I have ever seen do that. But we realize your custom permits you to use both hands." I replied, "Thank you Admiral but I prefer to sit on my left hand."

He invited me to eat with him and his staff. I was sitting next to him. You should have seen the faces of all his staff when I put my left hand under my butt after sitting down. The Admiral told all of the staff what I did that morning. They all laughed, but they really appreciated the gesture. They must have told everyone they knew about this crazy American captain from the American Navy ship honoring their custom. This custom arose because in their country you use your right hand to eat with and your left hand to wipe your butt.

The Admiral asked me if there was anything he could help me with while I was in port. I told him that I had some very important parts for our engine that were tied up in customs. He pushed a button on his desk and an English-speaking aide came into the room. The two conversed in their own language. When we got outside

A Military Mustang

the Admiral's office, I asked the aide, "What did he say?" He told me to go down to customs and get the parts you need for your ship. He said don't come back until you get them. The young officer said to me, "Don't worry, Captain, my uncle runs the customs office and you will get your parts, I assure you." They even tried to give me a big radial engine that had been in customs for two years that was for an Air Force C130. I enjoyed my stay in Karachi.

Tampa Bay

On December 28, 1982, the USNS Wyman was leaving Tampa, Florida, from Ybor City to a sea buoy that was 42 miles away. The day started out very windy with light snow which was unusual for Florida. The pilot was taking us out and we were proceeding to the sea via rows of buoys with red and green lights. When we got halfway out the pilot told me that it was too rough outside the buoy to have the pilot boat pick him up. He was going to get off at an island and I would have to take the ship out myself.

He got off and I proceeded following the buoys. After making various turns ahead of our ship, I saw a tugboat with a tow, crossways in the channel with the tow up against the bank. I talked to the tugboat captain and he told me that he had to keep the tow against the bank. If he did anything else with the wind blowing crossways in the channel, he would set down on the left side of the channel and lose his tug and tow altogether.

I immediately placed my ship in a crabbing motion allowing me to pass the tug and tow without the wind causing me to sit down on the left side of the channel. This was a hairy operation. First, I was committed to do what I had to do to keep from going aground myself. I told the tugboat captain that I would be pointing my bow towards the tow, but not to worry because we would safely pass by him at the last minute.

He said to me, "Are you sure you know what you are doing?" I assured him I was an experienced captain. I spent many years in the Arctic dodging icebergs and bergy bits that could put a hole in your ship. We approached the tug with forty-knot winds, heading straight at it. The captain of the tow and my own crew were in doubt that my procedure would work. Just as we were coming up close, I had the man on the wheel come left and we set down towards the opposite bank. As soon as we passed the stern of the tug and were clear, I again went back to crabbing towards the right side of the channel and proceeded out to the sea buoy.

The tug captain came on the radio and told me that it was the finest piece of ship handling that he had ever seen. He was sorry that he was blocking most of the channel. He said he was going to ride out the storm even if it took all day. I complimented him on his commitment to his vessel and the maneuver he had to make. Once abeam of the sea buoy we were no longer in harm's way and proceeded on our way.

Chapter 24 – PROTOCOL & PROCEDURES

Chinese Salute

In the beginning of April 1979, I was the Commanding Officer of the USNS Wilkes. Our ship was going into the port of Colombo, Sri Lanka. As we were maneuvering the ship to the dock, I saw that we were going to pass a Chinese ship at anchor. Our crew was at the mast getting ready to lower the American flag. I asked the pilot to stop the ship so we could drift by slowly and give the captain of the ship from China a chance to lower his flag.

I was standing on the wing of the bridge and when we were slowly going by, I saluted the watch officer who quickly sent someone for the captain to stand next to him and two men running to the stern to lower their flag. I saluted him and held the salute until we went by his bow. It worked out perfectly, allowing us to show respect and giving the Chinese captain time to do the same. I thought it was a nice gesture and proper thing to do.

When we left port with the same pilot, he told me that he met the captain of the Chinese ship who said he was called to the bridge quickly to return my salute. Everything went well at that time. I thought it was correct for me to do, even though in 1951-1952 we were on the front lines trying to kill each other. That was a long time ago. It was 1979 and we were at peace.

Saved by the Bell

I was the Commanding Officer of a US Naval Survey Ship in the Atlantic Ocean and was working with a million-dollar piece of ocean equipment to hunt Russian submarines. It was a very important piece of equipment. We made two passes with the scientist trying unsuccessfully to pick up the unit. I had a great 1st Officer aboard who could handle the ship. Being a Diving Officer, I told him to maneuver the ship, I was going over the side to put a line on the buoy which was attached to the valuable instrument.

I went down to the main deck and proceeded to take a line from the ship to the buoy that was right alongside the middle of the ship. The wind and seas were coming from the starboard side, the same side as the buoy. When the line was ready to be attached to the buoy it disappeared under the ship. The ship commenced drifting away and one of the young seamen on the ship thought I was in trouble and jumped over the side. He meant well but now there were two of us in the water.

The First Officer did an excellent job maneuvering in a circle. The ship came around on the windward side so the ship could drift down on the two of us in the water. The line attached to the buoy was floating on the surface. All we had to do was put a ladder over the side with a boat hook, grab the line and bring it aboard to a block and winch, haul in the line, and bring the buoy aboard. On the other end of the line was our prize…. The million-dollar device that could find submerged Russian submarines.

A Military Mustang

A Navy Commander was aboard to observe our operation. He notified his Admiral in Washington that the captain left the ship while at sea to recover a priceless piece of equipment. All hell broke loose, but I was ahead of him by sending a message to Commander Gooden at his base, right away. He was the Commanding Officer of Military Sea Lift Command, meaning the Navy ran the ship we were on.

Captain John Lawrence (Jack, to everyone that knew him) was a young Captain in World War II and now the Port Captain of MSC and a good friend of mine. If you recall, he was the guy who told me years ago what it would be like when I wore the crown.

Captain Lawrence went to Commodore Gooden and told him that I didn't think twice about going into the water to retrieve the million-dollar piece of equipment. Commodore Gooden didn't know what to do. He said, "I don't know whether to give him a medal or bring him up on charges. Since the Navy Commander aboard the ship didn't go through me first, I am not doing anything to Captain Arens." As they say in boxing, "Saved by the bell."

Pollywogs to Shellbacks

I was captain of the USNS Towle. The ship was a Victory Ship built during World War II. It was used by the Navy for 22 years in the Arctic and Antarctic to bring supplies to the Air Force bases. They could land on wheels in the Arctic but in the Antarctic, they had to land on an ice shelf. I was the captain because of my Arctic experience.

The difference between the two bases was distinctive. The Arctic icebergs formed from the land and came out between mountains. Greenland is 837,000 square miles of ice, 10,000 feet thick. The weight of the ice pushed the middle of the land down. In the Antarctic the ice breaks off the Ross Ice Shelf which alone is the size of France and comes out from the land in Antarctica.

The ship was to leave the Seabee base in California, cross the equator, and make shellbacks out of pollywogs. This was an ancient tradition, performed by sailors for hundreds of years when crossing the equator. First the pollywogs had to go before King Neptune and the royal baby. The royal baby was the seaman that had the biggest belly on the ship. The pollywogs had to kiss the royal baby who was wearing a white sheet as a diaper. When we crossed the equator, we had to crawl 20 feet through chutes that contained rancid garbage and other unpleasant things that had been saved from the last port. It was really disgusting in the chute because of all the rotting garbage. As you crawled on your hands and knees through the chute, other seamen would be hitting the chute with boards. It was a humiliating experience for sure.

You had to carry your documents with you when you were in the chute or you would have to do it all over again. You can't just say you did it. No one on the ship would believe you. This is a real ritual and still continues. I became a shellback in 1947.

Helicopter Ride to Dry Valley

While we were in McMurdo Navy Station off-loading cargo and equipment, I met a Navy Helicopter Pilot. Knowing I was writing to school children, he asked if I would like to take a helicopter ride. Of course, I said that would be great.

Up we went. Our ride took us about 50 miles from the base. On the way back he landed on one of the dry valleys. He told me, "Nobody has been in this spot for a million years. Get out and pick up some stones to send to the school children. You can tell them what I just told you." He gave me a bag and I loaded it with the stones. We went back to McMurdo Base which is on the west end of the Ross Ice Shelf and I took them back to the ship. The pilot logged it as a training flight.

The school children loved it when I brought the stones and told them I collected them in a remote section in Antarctica. After I retired, I went to schools to talk to children from 1965 to 1987. I enjoyed spending time with the children and answering their questions about life in the military and the different places I have been.

Unrepping Underway

I was the Commanding Officer of the USNS Rigel in the Indian Ocean. The ship was attached to the fleet with a carrier group that had the USS America and the cruiser USS California along the starboard side. The large carrier on the port side was usually a supply ship. The Rigel was between them passing frozen food and dry food to each of them. The carrier was usually alongside for 8 hours.

The Rigel maintained a speed of 12 knots. The other ships adjusted their speed to keep up with the Rigel. They call this unrepping, where they pass all kinds of material. The seamen on our winch pays out and the other winch man takes up the slack. It takes great skill to keep the load on a pallet from hitting the water. At 12 knots you will lose the load if you are not working together. With teamwork, everything can be passed without a problem.

I sent two Navy caps on top of one of the pallets, one for the captain on the carrier and one for the cruiser captain. After a short time, the captain on the carrier called me on the phone and said, "Captain, would you look up above me on the upper deck?" It was the Admiral of the fleet pointing to his own bare head. "He wants a Navy hat, too." I said to the Captain that I would send the Admiral a hat on the next load. They shared pictures they took of our ship laying alongside their ship. Everyone was happy.

Unrepping is one of the most dangerous times at sea. This is when you are with the fleet so there can be an aircraft carrier which always approaches you on the port side. There has to be coordination among the ships. While this maneuver is taking place, the most dangerous place on the ship is the main deck where seas coming over the deck cause hazardous conditions. Coiled lines are fired out of a bucket with a rifle. The lines are shot to the ship alongside and are attached to a wire cable that

A Military Mustang

follows the shot line. The seamen on each ship must work together. A piece of cargo is picked up off the deck and the receiving ship picks up the slack as the load passes over the water. Each man keeps enough strain on the line to keep the load above the water. The winch men on each ship are highly trained seamen.

While all this is happening, the officers are overseeing from the bridge of each ship. The ships have a very experienced sailor at the wheel with an officer giving orders to the helmsman to keep the ships a certain distance apart. The captain of each ship is standing by at all times. The ship in the middle has a pee-can because no one can leave the bridge wings even to use the bathroom.

A good captain is always training the young officers to unrep safely, explaining the dangers when one of the ships gets too close to the one in the middle. And instructing how to hand the situation. The officer in charge of the ship moving in too closely must give orders of small movements to the man on the wheel. If you give too much and too fast, the bow of your ship will move away. The stern will swing in and hit the ship in the middle which is the ship carrying supplies for both ships.

Another serious problem is when a wire breaks apart. The steel cable is under tension. If it breaks, it can fly up and wrap around seamen working on deck. The key is strict coordination. Besides the officers on the bridge you always have a rig bosun and a top bosun running everything.

Another part of unrepping is the ships receiving supplies have to maintain a certain speed to stay with the ship in the center which is the one giving the fleet supplies. If you have an aircraft carrier receiving supplies, they have helicopters flying aft of the ships to pick up any seamen washed overboard or loads that hit the water. If you are passing jet engines for the fighter jets on board, they are in a pod. If they get dropped into the sea, the helicopters can pick them up and return all cargo back to their ship.

Training is what makes a ship The Best. The engine room personnel have to be ready for instant bells on the ships receiving. The re-supply ship in the middle must maintain 12 knots at all times. Unrepping is an art that's been practiced for many years on ships before World War I.

I was pleased with everyone working together. One of my most outstanding bosuns was Jason Dill, he was a former Marine Corps Semper Fi. He was one of the few men who always walked like he was at attention and would just glide along like he was floating on air. All the men respected him. He kept his own records on the men. If someone complained that he wasn't fair, he would report to the Captain's Room with his book of notes. He always won his case because he had a record of his interactions with the men. He was first class all the way.

A Military Mustang

Chapter 25 – FAMOUS MEN & WOMEN

Steve McQueen

While I was Commanding Officer of the USNS Towle, we were sitting in Port Hueneme, California. Around the end of December of 1979, one of the men brought aboard to work on our electronics told me I should visit Santa Paula Airport, where pilots do stunt flying for movies. Not only that, he said there were a lot of old planes, including double-wing aircraft. I decided to drive down and look the place over.

When I arrived at the airport, I went to the front of a hanger where there was a long nylon rope extending across the front to keep people away from the planes. I looked in and saw two aircraft, a yellow Stearman that was used to train World War II pilots and a big double-wing mail plane. I was surprised to see there were also about thirty-five motorcycles sitting in a circle with a German bike in front. A man inside was talking to two men. He looked up and when he saw me, he asked if I would like to come in and look around. The other men left, and he started to walk towards me. When he got closer, he looked familiar. As he got up to me, I said, "Are you Steve McQueen?" He said, "Yes, but just call me Steve."

Curious, I asked him about the German bike in front. He answered, "It is from the movie, *The Great Escape.*" He told me that he worked on all the bikes sitting there. He owned a ranch a short distance from the airport which made it easy to spend a lot of time at the hangar. I noticed the shelves on the hanger walls contained children's toys.

When I asked about them, Steve gave me a tour of his children's toys and said, "Most people would not believe I collect toys, but I have enjoyed collecting them for many years." He reached up on one of the shelves and gave me a 1929 racecar with a tailfin on it that I kept for a few years. A cousin of mine from Kentucky had a nice collection himself. Unfortunately, there was a fire and he lost a lot of his collection. So, I gave him the racecar with an affidavit saying it had belonged to Steve McQueen. I was happy to know that people would get to see the car and know where it came from.

Steve and I got along even better when I told him I was an old Harley Hog rider. We talked about the movies and I mentioned I had watched *War Lovers,* in which he played a pilot. He explained to me that he learned to fly the Stearman bi-plane a year before for that movie.

I told him that I liked the movie, *Enemy of the People* in which he played a doctor with a beard. The movie was about a town in Norway where he found the water was contaminated. The town businessmen didn't want it brought out because the tourists would stay away. He immediately questioned me, "Where did you see this movie?" I said, "On the ship." He said that he was told that movie had been shelved and that he was going to call

A Military Mustang

his lawyers and get the money rightfully owed to him because the movie was being shown.

Steve asked me if he could visit my ship. We were heading to New Zealand, then to Antarctica to Captain Scott's Station. We were taking equipment for the personnel who would be there for the winter months, but we were not leaving right away. So, I said yes. He introduced me to Barbara Minty who he was to marry in a couple of weeks. He asked Barbara if she would like to visit my ship for the evening meal. She said, "We can't go tonight but we could go to the ship in a couple of days, if it will still be here."

I told them we would be here, and I asked him what he would like to have for dinner. Without hesitation he said, "Roast beef, mashed potatoes, vegetables, and apple pie." I said, "OK."

I told him I would pick him and Barbara up in two days at about 2:00 p.m. He asked if he could bring another lady. I was surprised to find he wanted to bring a nurse because he looked pretty good to me. I went back to the ship and told the Chief Steward that Steve McQueen, the movie star, was coming to the ship for supper in a couple of days; would he make roast beef and all the trimmings. Before I drove back to the airport to pick up Steve, I asked the Ship's Purser and the Navy weatherman assigned to our ship to come along when I went to pick up him up. They did not believe we were going to get Steve McQueen and bring him to the ship.

When we were approaching the airport, I saw the yellow biplane coming in for a landing. I said, "There he is now." They still didn't believe it. When we came up to the side of the hanger and parked the car, they said, "Why don't you tell us the truth? There is no Steve McQueen here." We waited in the hanger as Steve was bringing in the aircraft. It had an open cockpit with two seats, one behind the other. He stood up in the cockpit, took off his leather flying helmet and said, "Hi, Captain, are you ready to go to the ship?" The two guys were so surprised, saying, "It really is Steve McQueen!" I said, "Oh, ye of little faith."

Steve followed me in his pickup truck with the two women who had driven to the airport together. At the dock the stevedores, dock hands, were loading the ship. I told Steve I wanted to check the draft to see how they were doing. Once the stevedores knew it was really Steve McQueen, they were hollering, "Hey, Steve Baby!" and lining the rails with the crew. We walked aboard and got ready to eat the evening meal. Before he left, he told me that when the ship returned, he would take me up in the yellow biplane.

When we returned on March 4[th,] I was told Steve wasn't there and he would be gone for a while. I found out later that he died of cancer. He had spent time in Mexico trying to save his life, but it was not to happen. I will always have good memories of Steve, the actor and Steve, the man. It was such a pleasure to meet him.

Stunt Planes

Before leaving port, I visited the Santa Paula Airport where I had met Steve McQueen. I saw a hangar

A Military Mustang

with four Pitt Special aircraft in it. They had high speed engines and three bladed propellers. Later, I walked into the hangar to see about getting a ride on one of the stunt planes. A pilot said he was going up in about a half hour and because I was the Captain of a large Navy victory ship, I would not have to pay.

He fixed me up with a World War II style helmet with goggles and an intercom so we could talk to each other. Well, he gave me the ride of my life! We did loops, split SSS and hammerhead stalls—going up until he would lose speed and fall backwards, through his own smoke. He kept describing his flight maneuvers to me all the while. He said, "You're not going to heave in my aircraft, are you Captain?" I said, "You can't make me sick. I was an Army Ranger, 187th Airborne Regimental Team, and a Navy diver trained by Naval Seals. Didn't you ask me that on the ground?" He replied, "You don't know what it is like to clean up the cockpit. They let everyone fly, but I'm the one who has to clean it up."

We stayed up for 48 minutes and I enjoyed every second of it. He finished the ride with a perfect landing. I offered to bring him aboard the ship to have a meal, but he said he was going to meet his wife. He had promised to take her out to dinner. I was really happy to go up in a Pitt Special. It has a small but powerful engine that looked beautiful on the ground and in the air.

William F. Buckley

In 1982 our ship, the USNS Wyman, was sitting in the Azores in the Atlantic Ocean. We were secured at the pier with water and electric. While up on the wheelhouse I overheard a conversation taking place on a beautiful white yacht It was about 70 feet long and worth $650,000. The captain was talking to the Harbor Master asking where he was supposed to dock. After he gave his name, he was told there was no space for his yacht. Overhearing his name, I asked him, "Are you William F. Buckley, the Editor of *The National Review*?" He replied, "Yes, I am."

I said, "I am Captain John W. Arens on the white naval ship just about 3 points off your port bow. I am tied up to the pier." Bill answered a little confused, wondering why I was giving him all that information, "I do see you." I said, "How would you like to bring your yacht alongside my ship? If you do, I will be able give you water and electricity." He immediately shouted back, "Great! I will be right over."

We put fenders over the side and the Sealestial pulled right alongside. We hooked up the electric and water and he was as snug as a bug in a rug. He said, "I can't thank you enough, captain. I said, "No problem, Bill." When I visited the Sealestial, I saw this beautiful chair. I told Bill that my crew was going to steal it that night and replace it with a papier-mâché model. Before we left, Bill gave me the chair as a thanks for letting him hook up for electric and water. Whenever I wrote Bill letters, I always reminded him how comfortable his chair was on my ship. He included part of this story in his book *Atlantic on High*.

A Military Mustang

He asked that if they ever make a movie of my life, could he write the script? It surely was nice of him to make an offer like that. Unfortunately, he passed away before this was accomplished. I always wondered what movie star would have played my part. My choice would be John Schneider.

Queen Elizabeth

From August 28, 1981 to March of 1982, I was Commanding Officer of the USNS Rigel during naval operations re-supplying the fleet in the Indian Ocean. The Rigel carried fresh and frozen vegetables, plus parts needed for ships in the fleet. When we arrived in Singapore, our ship was docked behind the Queen's yacht, Britannia. Prince Charles and Princess Diana had just gotten off the ship to continue on their honeymoon. Before docking, they used British Navy divers to check out the pier to make sure everything was clear on the dock and in the water near the dock.

A sailor came to our gangway inviting me to board the Queen's yacht for a tour. The Chief Engineer and I arrived wearing our best uniforms. We were escorted to the officer wardroom where we talked to their officers. We took a tour of the yacht which was immaculate top to bottom. They manned the top deck but steered one deck below with a seaman at the wheel.

We went to the stern of the ship on the starboard side where there was a large room for the Queen and a smaller one on the port side for Prince Consort Phillip. The carpets were a light beige. A seaman was vacuuming with a cloth wrapped around his shoes so he would not track in dirt while cleaning. There was an area in front of her room that was used for showing movies. The room could hold about 20 people.

We proceeded to the engine room and could see it was immaculate just like everything else. If you did a white glove inspection, you would not find a speck of dirt. I think the crew was pleased when I told them they should be proud of the engine room.

After that I was escorted to the crew's quarters where the Chief Engineer and I sat with the crew members. I was impressed as one of them talked about Princess Diana. He was telling us that he was scraping the deck when she stopped to talk with him. She asked him how things were at home and about his family. He politely told her by order of higher-ups that he was not to carry on a conversation with her.

Princess Diana went to the captain of the ship and told him, "Whenever I stop on deck and talk to one of the crew members for ten minutes, half an hour, or whatever, I expect him to carry on a conversation with me." I could see why the crew loved her. They also showed us the ship's logbook where kings and queens had signed their names when visiting the ship. Names of royalty from all around the world were in the logbook.

When I arrived back in the States, I visited Green Hill Beach at my daughter's residence in Rhode Island. I was walking on the road with a gentleman who lived in a beautiful mansion nearby. We stopped by his house to meet his wife. I looked down at her beige wall-to-wall

A Military Mustang

carpet and told her that it was the same color as the rugs that were on Queen Elizabeth of England's yacht. She was pleased to hear that. I'm sure she told all of her friends. I think I made her day.

King Frederik IX of Denmark

In 1968 our ship, the USNS Redbud was in Kulusuk, East Greenland. The Redbud which boasted a Navy Diving Team of four personnel was responsible for re-supplying Air Force Bases. This was my fourth year as Diving Officer. The ship carried 4,000 feet of 4" hose to re-supply Kulusuk. The hose had to cross a small island to get to the mainland where the tanks were stored. When we finished, we picked up the hose and stored it in the hold of the ship. After completing the operation, a tanker would come in and pump off fuel for the base.

While we were anchored in the harbor there was a Norwegian Navy destroyer anchored nearby. King Frederik of Denmark and his Queen were making a visit to the ship. There was a lot of ice in the harbor. Our ship was sitting at anchor and the destroyer was in a clear area. The King's gig was leaving and came right at us. When they were close by, I blew the ship's whistle three times which is an international salute. The King and Queen stood up and rendered a hand salute. The Queen waved at us looking down from the bridge. I returned the hand salute. When the King arrived ashore, he told the manager of the base, "That is a real sailor." He was very pleased with the respect we showed by saluting.

When the King was a young prince his father put him in the Navy and told the officers to treat him just like the other cadets; he was to get no favors. One of the stories he told me was that one of the men kicked him in the ass. Years later when he became King, he was drinking with the crew. He pointed to the bosun and told the young sailors, "When I was young like you, the chief who is sitting there kicked me in the ass." The chief almost choked on his drink and said, "I had orders to do that, my King." The King laughed and held no malice. Whenever he came aboard the Navy ships he loved to talk and visit with the crew. I enjoyed spending time with the King.

Chapter 26 – HELPING HAND

Pickle Jars

I made it a point to save all glass jars, especially large pickle jars. Other MSC ships and various restaurants in New York, Staten Island, Bayonne, and Norfolk would save their large pickle jars for me as well. We would clean them up and give them to poor women when we were in port. The women loved them. They used them to hold clean water in Sri Lanka, India, Seychelles Islands, Egypt and other countries. Many times, we would come into port and the women remembered the ship and showed up to get the valued pickle jars.

At one of the ports in East Africa I met one of the local soldiers. He invited me to his house. A driver took me and the soldier to his house. What I did not know was they had dirt floors. When we arrived at the house, his wife had made a design in the floor with her broom. When I walked in, she followed me with her broom and made a design in my footsteps. I told her what a beautiful floor she had. She was very happy that I told her how nice the floor looked.

Karate Teacher

Our ship was traveling in the Caribbean Sea. One of my seamen aboard the ship was teaching Karate to a few men, including the captain of the ship. He was very good at what he did and always told everyone that he taught to use Karate only if you are attacked. At no other time should it be used because anything you do could be lethal. While we were in port on one of the American owned islands, I had a police officer come to the ship at the pier. When I reported to the gangway, he told me that they had a seaman from my ship in jail for attacking one of their people from the island. The man was in bad shape at the hospital with a fractured skull.

I immediately went to the jail with the officer only to find out it was our Karate teacher. I told the chief that my man would never attack anyone as he would only use Karate if he was attacked. When I talked to the Karate teacher, I listened to his story. He told me, "I wasn't doing anything, but this guy comes at me with a long chain and demanded my wallet. When I saw how he was holding the chain, I knew I would be able to defend myself. As soon as he swung the chain, I did a round house kick that caught him right on the side of the head and dropped him like a rag doll.

The next morning the headlines read: AMERICAN SEAMAN ATTACKS ONE OF OUR PEOPLE! It was big news. After the paper came out with the story, a black lady came to the police station to say that she was a witness and described the attacker as a Rastafarian with long black hair. The guy in the hospital fit the description. Well, the chief of police let my guy out and apologized to me for arresting my seaman. The next thing he said was, "I would like to have him teach the officers his moves." Of course, we were always happy to help local law enforcement.

Food for the Base

The USNS Rigel had orders to dump all the unused food that was on the ship into the sea on the way back to the United States from Roto, Spain. Having grown up in the depression, I could never see wasting food when it could be given away. The base we were leaving was a very large Navy base. So, I had the crew bring all the food up on deck and put the word out that we would be giving it away to anyone stationed at the base. The enlisted wives started coming right away. The officers were not going to touch it, but their wives saw the enlisted wives eagerly grabbing all the food and joined in.

There were lots of vegetables, frozen meat, and fruits. The apples were going fast as all the women wanted to make apple pies. One of the wives brought me a pie she made with the apples. It was very good. She told me she had her officer husband peeling the apples. The nicest part was that it was the best-kept secret in the Navy. Nobody complained, and I never heard a word from the base when we got home. They asked me if I got rid of all the food and I answered, "Yes, I did."

Clothes for the Orphanage

My Aunt Pauline Bethel knew a Priest in Corinto, Nicaragua. She had collected 86 boxes of clothing to give to him. When he came down to the ship to accept the clothes, he told me the customs officer said he would take half for himself. I asked him to take me to the customs officer. When we arrived, I met the officer and explained, "I am the captain of a US Navy ship and I was going to give 86 boxes of clothes to an orphanage in another country, but that customs officer wanted to take half. Since I knew you were an honorable man, I decided to come to your country instead. I knew you would see that the Priest would get all 86 boxes of clothing."

He answered back, "Captain, I will put guards on the truck which will deliver it to the Priest in the small village up in the mountains to make sure no one takes any of the clothes." I told him, "Thank you and I will let the Navy Admiral know what a good deed you did." When we left the customs office, the Priest told me, "Captain, you should be in politics when you retire. You handled the situation perfectly, allowing me to get all the clothing and the customs official to get all the credit for it."

A Military Mustang

Chapter 27 – SHIPS

Captain of Two Ships

The USNS Hayes was built as a twin-hull ship, which is like a catamaran with an engine in each hull. There was no way to get to the other side from below, but you had to come up on top to get from one hull to the other. When I reported to the base for my new assignment, Captain Lawrence told me, "We think so much of you, we are going to give you two ships to run." He chuckled as he told me it was a twin-hull. It was quite a ship. Captain Couch died on the ship and I was going to take over. I had been his first officer on the Redbud when it was working in the Arctic.

On the USNS Hayes we worked with marine scientists off the coast of the United States that were experts on earthquakes. The scientists' job was to predict earthquakes. They installed deep sea buoys with instruments that collected seismic data which would be relayed back to the scientists. This information was very valuable and could save many lives.

One time the USNS Hayes was anchored in a Rhode Island harbor near the naval school. A young man with two young girls took his speed boat between our two hulls. When they got in the middle, they were hollering just to hear their echo. We could hear them. They came in from the back going towards the front pretty fast. That wasn't a smart thing to do as they could have hit any hanging equipment in the middle of the ship.

German Battleship Tirpitz

Our ship was in Bergen, Norway, I made a courtesy visit to the Admiral of the Norwegian Navy. He told me about the battleship, Tirpitz. It was named after Grand Admiral Alfred Von Tirpitz. He was the architect of the German Navy. The Tirpitz was commissioned in 1941. She was the sister ship to the Bismarck that was sank by the British. She was hit in the screw (propeller) by a bi-wing plane causing her to run in circles. Disabled, the British battleship then blew her apart with gunfire and she sank to the bottom of the sea. Over a thousand men went down with Bismarck that day.

The Tirpitz was sent to Trondheim, Norway and moored next to a cliff which protected her from the southwest. She was camouflaged with trees that were cut and placed aboard the ship. The ship was eventually caught at Hakoya Island outside Tromso. The final attack occurred on November 12. As the list (the leaning to one side of a damaged ship) increased to 40 degrees, a large explosion rocked the turret, rolling her over and burying her in the mud. Over a thousand men were lost. Ludovic Kennedy wrote in his history of the vessel that she lived an "invalid's life and died a cripple."

The Admiral also told me a story that was little-known outside of Norway. When the Germans took over Norway during WWII, the only ones killed were individuals who resisted or who had shot German soldiers. The Germans felt the Norwegians were traitors when the sailors escaped, leaving Norway to join the ships at war with Germany. If they caught the Norwegians on a ship when they took it over, they were taken back to a fjord in Norway where weights were put on their feet and they were dropped into the fjord which was 600 feet deep while they were still alive. The Admiral told me, "The fjord is their tomb and we honor them."

Arriving in Port and Docking Ship

When docking a ship, the crew must be well trained as young seamen and must learn the ropes on how to bring a ship up to a pier safely. Accidents can happen quickly. The first line that goes out is the forward spring line which is the most important because this is the line that controls the forward motion as you approach the dock. Most of the time there is a docking pilot who knows the area but sometimes you have to do it yourself.

In the Arctic and Antarctic most ships can be controlled at about 15 RPM when the spring line goes out to a heaving line. We are talking about throwing the heaving line attached to a big nylon line. The men on the dock will pull the big line onto the pier so the stevedores can put the "eye" (noose) on a ballard. That will stop the forward motion of the ship. You have to be prepared to not have many turns on the bit. You can now continue slacking which will prevent the line from smoking and burning. That will stop the forward motion of the ship as soon as possible. The procedure is being done aft, at the same time.

Bow and stern lines are thrown out to bring the ship snug against the dock. You have a windlass on the bow of the ship to put an anchor down or up, but you also have capstan to wrap line around to keep the ship safe and snug to the dock. Sometimes you wish you had eyes in the back of your head.

Chapter 28 – CHALLENGES AND AWARDS

Safety Award

In September of 1982 the officers and crew of USNS Rigel received top honors when they were named winners of the ship Safety Achievement Award for rescuing a flooded vessel under adverse weather conditions in the Indian Ocean. This maritime Safe Achievement Award is presented annually by the American Institute of Merchant Shipping and the Marine Section of the National Safety Council.

I, as Master of the Rigel at the time of her achievement, accepted the award on behalf of the crew. The ceremonies were held at Military Sealift Command headquartered in Washington, DC. Rear Admiral William M. Benkert USCG (Retired), the AIMS president, cited the Rigel crew for their outstanding feat of seamanship and damage control expertise. The crew of the Rigel worked for more than 80 hours in winds that rose to 25 knots and seas that were up to 10 feet to help salvage a Thai vessel that was flooded due to a crack in her hull. They were able to temporarily repair the crack in the vessel, the SS Bahar Alsiam, and to dewater the holds so that the ship's means of propulsion was restored. After three days of hard work by the crew, the Thai vessel was able to sail under her own power to Sri Lanka. Admiral Benkert said, "There was no loss of life or ship. It was just a helluva a good job of seamanship, Captain."

Accepting the award, I noted that the Rigel crew did the work required to salvage the ship. "I just took our ship alongside while my men went aboard and made the repairs." I attributed the feat of seamanship to the Navy practice of insisting that MSC crews train constantly. Our crew had been well trained to meet the emergency that resulted in the saving of the Thai ship.

The Rigel also received a pennant bearing the Green Cross of Safety, presented by Captain Peter J. Cronk, (USCG) on behalf of the National Safety Council's Marine Section. I promised to fly the pennant from the ship's yardarm as long as it lasted. The Admiral stepped up behind me and put his hand on my shoulder and said, "What you said about the crew is what I would have said." He made my day!

Hurricane

I was Commanding Officer on the USNS Wyman in 1982. A hurricane was forming south of our ship. It was observed due south. Our ship was directly east of the hurricane when it veered and began heading towards us. The waves were picking up height. So, I ordered the officers on the bridge to keep the wind and seas on the port bow and slow the ship down to prevent damage to the ship. The ship was rolling up to 35 degrees. I ordered the Chief Steward to serve only sandwiches and cold food until the danger was over. With the ship's heavy rolling, boiling water and hot grease could spill burning the cooks.

A Military Mustang

I was in my quarters and decided to go up to the bridge to see how the watch officer was doing. I told him and the young redheaded seaman on the wheel that I had complete confidence in them. In fact, I was going down to my quarters and finish reading my book. After I left, the young seaman laughed, saying to the watch officer, "What the Captain didn't tell you was that he was going down to read the Bible."

The ship continued to a safe port to wait till the hurricane moved on. Fortunately, the ship suffered minimal damage. After the storm we were able to continue safely into the Azores.

Putting My Foot in My Mouth

An admiral was visiting our ship, the USNS Harkness, in 1983. His helicopter landed on the ship's helipad aft because we carried our own helicopters with Navy pilots and crew aboard. I met him on deck with our side boys and one female side girl who I personally asked to be with the group. The young lady who was a Naval Seaman was delighted when I personally invited her to be the first woman side girl along with the side boys who are always present when a high-ranking dignitary comes aboard a ship. This is an old custom followed since the early days in the Navy.

She said to me, "Gee, Captain, no one has ever asked me to be a side girl." At the time, she was one of the first female side girls at sea and I was glad it was on the ship I commanded. While I tell this story, I must admit I didn't always get it right.

The Admiral came to my quarters and I asked him to have a cup of coffee. He said, "Yes." I went down to the mess hall myself because everyone else was busy. When I arrived at the mess hall, I saw a woman with full flight gear drinking coffee. So, dummy me, I said to her, "What is it like to run that winch aboard the helicopter?" She looked me in the eye and said with a straight face, "How the hell do I know. I'm the pilot!"

Without flinching one bit, I put my shoe on the bench, unlaced it, and pretended to take a bite out of it. This is what any smart commanding officer would do after making a gaff like that. She started laughing. I could just see her telling that story the rest of her life and enjoying every minute of it.

Mine Sweeping

I was Captain of the USNS Harkness. The ship was laying at anchor close to the southern part of Port Suez Canal. Muammar Gaddafi of Libya had his henchmen drop mines in the area near where my ship was anchored. We were in the canal passageway headed towards the Red Sea. I heard a loud explosion and immediately called the engine room. I was informed that we were not hit, but it must have hit somewhere near. As I neared the bridge, I saw the watch officer scanning the area with binoculars. He saw a vessel bearing a group of men on hajj (pilgrims going to Mecca in Saudi Arabia). The vessel was listing, and it wasn't long before an Egyptian navy boat arrived alongside our ship.

A Military Mustang

The officer wanted to see my logbook. The Minister of Defense thought my Navy helicopter dropped a mine in the water causing the hajj ship's demise. I refused to show him my logbook but told him as the American Captain of the Navy ship, Harkness, I could assure him that my helicopter did not fly that morning. Therefore, it did not drop a mine on the vessel. As the Commanding Officer of a Navy ship, I immediately sent a FLASH message to Military Sealift Command informing Washington, DC and other commands of what was going on. (Demanding to see the logbook was a direct affront against the United States of America.)

All hell broke loose after my message. They sent an EOD (Explosive Ordinance Disposal) team from Sigonella, Italy, over to our ship along with a large ship with an opening in the back to bring small ships inside. I remained the On-Scene-Commander until the larger ship arrived. England was sending three mine sweepers and other countries were involved with mine sweeping. Making matters more volatile, the EOD Team was run by a female officer which really annoyed the Egyptians. They wanted me to send her back and get a male officer. I told them that I had no authority to do that. If I sent her back the women senators and congressmen would have me relieved of my command.

I told her that I would send her into the beach area during the day. She would come back to the ship at night and then go back the next morning for safety reasons. That also put her first-class diver in charge at night. We became a mine sweeper with our ship dragging for mine detection. When the big Navy ship arrived, it took command.

The helicopter pilots knew they did not fly. Our ship officers knew our helicopter didn't fly and the engine room knew they didn't fly. The Egyptian government was the only one claiming our helicopter was the one that dropped the mine. After three months, I received a letter from the Head Admiral praising me, saying I did the right thing and performed throughout the whole operation as a U.S. commanding officer should. The operation lasted 30 days.

In 2011 it was the same Gaddafi, President of Libya who had been giving us a hard time for 18 years. Finally, his own people shot him in the head and dragged him through the streets.

Offending the General

General Robert Kingston was arriving aboard the USNS Harkness, specifically because he did not receive an important message that went to Washington Chiefs of Staff. He had two Air Force Generals and my Admiral with him. I had a Navy officer aboard with a group of Navy personnel who handled all messages. However, the Captain is responsible for everything aboard his ship. I was tipped off that he had fire in his eyes and was going to hang me and the Admiral for gross negligence. I was prepared for the worst.

When the General arrived on the stern of our ship, he passed through the hangar deck. I was standing at attention farther up the deck to meet him. I was wearing my six rows of ribbons, CIB on top, Ranger Jump Wings, Gold Badge of a Navy Diving Officer, and Gold Navy Jump Wings. He was with two Air Force Three-Star Generals. My Two-Star Admiral was following the group.

Remember, he was going to hang me. When he got in front of me, I saluted him, and he saluted me back. While he was still holding his salute, he said to me, "What are you doing with that CIB on your uniform and Ranger Jump Wings?" I quickly said, "General, I was an Airborne Ranger, 3rd Company, in the Korean War and also a Sergeant in the 187th Regimental Combat Team, both on the front lines. Trained by Navy Seals, I put ten years in the Arctic re-supplying Air Force Bases in Greenland." With that, the Air Force Three-Star General leaned out and gave me a thumb up, meaning, "You're in!"

The General then put his hand over my ribbons like a blessing, and said, "An Army Sergeant running a Navy ship! I would have never believed it." I immediately said, "General, we are holding up everybody. Let's go to my room." He answered, "Yes, yes, let's go." I saluted the two Generals and my Admiral because he was in the back and didn't hear or know what was going on up front, he looked glum.

When we got to my room, General Kingston was sitting close to me and he turned to the two Air Force Generals and said, "Can you imagine that, an Army Sergeant is running this Navy ship." My Admiral was sitting more to the back waiting for the axe to fall.

Everyone was drinking coffee but me. I drank tea. I explained to the two Air Force Generals that I was a Navy diver trained by Navy Seals and spent ten years (1965-1975) in the Arctic bringing in ships and tankers containing JP fuel for fighters and aircraft to Air Force bases in Goose Bay, Labrador; Sondestrom, Greenland; and Thule, Greenland.

At that time, I turned to General Kingston and said, "General, let's get to the point of what you really want to hear from me." He looked me right in my eyes and said, "It won't happen again, will it?" I said, "No, Sir, it won't." All was forgiven. My Admiral up to that time was saying nothing. He just slumped in his chair with relief when General Kingston said to him, "I'm satisfied, Admiral."

Everyone was asking me questions about diving in the icy water at the Air Force Bases. Later I heard about General Kingston bragging about this Airborne Ranger and 187th Airborne Sergeant running a Navy ship. It was much later at Fort Benning, Georgia, when I saw General Kingston's picture on the Army Ranger Hall of Fame that I understood why he changed his approach.

Dismantling Bombs

I was Captain of the USNS Harkness during the time that mines were being dropped by Gaddafi. He had his people drop mines to cause problems during the transit of the Suez Canal. We heard the explosion thinking it was our ship, but it wasn't. I was docked at a pier when I got a visit from the captain in charge of the

A Military Mustang

mine sweepers. When we finished our business related to his duties, he told me a very interesting story about his former time in the British Navy during the Falkland War with Argentina. A pilot from there dropped a bomb on the British ship, Sir Galahad. It went through three decks without exploding and he, being a demolition expert, had to go down to where the bomb lay and dismantle it so it wouldn't explode and destroy the ship. When he got back to England he went before Queen Elizabeth to receive the second highest award. (The highest is the Victoria Cross.) Before the ceremony he was told not to talk to the Queen. When she presented him with the medal, he spoke up anyway saying, "Your Majesty, your father presented this medal to my father who was a Spit Fire Pilot in World War II." The Queen answered him, "Thank you for telling me about your father. The captain brought his violin with him aboard my ship and played beautiful music for me. I always wondered how far up he went in the British Navy."

Dumpster Toss

Life at sea can be monotonous at times. Throwing in some fun and games helps to brighten the day. The USNS Redstone was on duty off the coast of Florida. It was never off the coast more than a few miles to work with submarines. During that time, we ended up with a lot of trash that was loaded on the deck along the rail by my direct order so that it could be unloaded when we arrived back in port. After docking, I contacted the Navy officer running the pier and personally asked a chief petty officer to get a forklift and put three dumpsters just forward of the gangway where all the trash was placed on deck.

Instead of having the steward department carry it all down the gangway, we had a contest as to who could drop the most bags into the dumpsters without missing the hole. The men had a great time dropping the bags accurately while dumping the trash.

Chapter 29 – OBSTACLES

Mine Hunting Operation

The Secretary of the Navy took pleasure in presenting the Harkness the Meritorious Unit Commendation to Task Force 155. Captain John W. Arens, Commanding Officer of USNS Harkness with the Oceanographic Unit Five on board embarked with the EOD Team from Siganella, Italy, and checked for underwater mines dropped by ships under orders from Gaddafi of Libya. The Task Force also included Egypt, The United Kingdom, France, Italy and The Netherlands.

The Harkness set sail to an area about 90 miles off Rasshukhayr, a port located on the Egyptian side of the Gulf of Suez. The Harkness was 393 feet long with a high freeboard which caused her to have a list if the wind came from different directions. The Harkness was in charge of the operation until the USS Shreveport arrived on scene on August 16th. The Harkness was running low on fuel. So, I worked out a plan with the CO of the Shreveport, Captain Ianucci to pass a fuel line between the two ships while moored stern to stern. This plan involved the Harkness slipping in behind by approaching bow to bow, run alongside and maneuvering in behind the Shreveport while passing stern lines to the Harkness. The Harkness received 16,780 gallons of fuel from the anchored Shreveport.

Finding no mines, the operations were secured, and the Harkness went back to regular operations. On October 2nd I received a letter from Admiral William H. Rowden of the Commander of Sealift Command in Washington, DC for a job well done.

Rest and Recuperation

Our ship the USNS Harkness was coming into Piraeus for R & R (Rest and Recuperation). A famous battle took place here in 480-479 BC. Xerxes, the Persian king was coming between the islands with his long boats with many oarsmen using long oars and Persian warriors. Oarsmen were usually slaves. The Greeks were waiting behind the islands. The Persians could only fight one galley at a time and the Greeks knew it. They would plow the Persian galley, one at a time, with rams located on the bow of the galley and destroy them. Xerxes stood on a hill watching the whole scene as his sailors were being slaughtered. Many jumped into the water but could not swim and drowned. The ones that could make it to the beach were killed by order of the Persian king.

We were the only American Navy ship anchored in the port because they had no room at the main dock. The problem with being anchored is the lack of electricity and a fresh water supply. As in other ports, I arranged to make a call on the Greek Admiral. Arriving at the Admiral's office in full dress whites, I was ushered into his office. The Admiral was happy to see an American Captain. He asked me about my various ribbons and was amazed that I was a diver in the Arctic and Antarctic with

28-degree water. He was also impressed seeing my Gold Wings for parachute jumping in the Korean War.

After talking with him for quite a while, he said, "Is there anything I can do to make your stay happy here?" I answered him, "Well, Admiral, I am at anchor in your harbor because they do not have any space at the pier." He looked back at me and said, "Well you don't have a problem now." I looked confused and he said, "I'm going to have you dock at the Navy base so you will have water and electricity." I answered, "That would be great Admiral. I will call my base and tell my Admiral's Aide about your generosity."

The crew were provided with special Navy buses to take them on a tour of the Acropolis. I had a private tour in the Admiral's personal car, and I was treated like royalty. We all had a great stay in port. Greece was a wonderful country for touring old historical sites and learning about its fascinating history.

Sister Rita and Mrs. Smith

I was writing to the children of two schools for many years: John Long School in Grafton, Wisconsin, and St. Mary's in Lomira, Wisconsin, about 30 miles apart. Mrs. Smith was the 5th grade teacher at John Long and Sister Rita was the 5th grade teacher at St. Mary's. I did not realize the two teachers were related to each other. It was funny that with all the different schools I was connected with, I was writing to two teachers who were actually sisters. The ship's crew would donate money to buy things in different countries and we would ship them to the schools. We sent artifacts after my ship made the same trek as Christopher Columbus' ships. John Long School received all the information and shared it with other schools. One day as the teachers were discussing the letters they each received from a Captain, they realized the same Captain was sending the letters to each of them.

In 1986, I was captain of the USNS Redstone and writing to Mrs. Smith's 5th grade class. The ship worked with NASA and the space program. Once the Navy and NASA found out that I was writing to the school children, I was provided with many items pertaining to the shuttle flight program to send to the school. The school was flooded with many interesting pictures, emblems, and patches.

The Space Shuttle Challenger

On January 28, 1986, the space shuttle, Challenger, was scheduled to go into space lifting off from Kennedy Space Center. The children were watching it on television mainly because of Christa McAuliffe, the teacher who was aboard. As many of you remember, seventy-three seconds after liftoff there was a large explosion, and all lives were lost. I knew the class saw it. I immediately called the school and talked to Mrs. Smith. I knew it was devastating for the young ones.

Not too long after the disaster I received a call from a colonel in the Air Force at NASA. He asked me if I had any equipment on that day because he knew we had the *Red* button that could destroy the booster tanks on each side of the main red tank that propelled the shuttle into space. The safety feature was for use after the booster tanks dropped off. If they went off course on the way down, the missiles could destroy them before they could do any damage.

I explained to him that nothing was on below decks and scientists and technicians were not aboard because we were doing engine maintenance. I also told him that it takes a Colonel in the Air Force to push the button. He said, "Thank you, Captain, I will explain all of this to the General." It wasn't till after the investigation that we knew the "O" ring seal was found to be faulty, causing the explosion.

Cape Canaveral

The Redstone was called "The Space Ship," because it was used from 1969 to the present time with the Space program. The ship made history in 1985 when the Range Sentinel had an engine breakdown and equipment was transferred from the Range Sentinel to the USNS Redstone for a missile shot into space. The USNS Redstone was laying by at the at the Cape Canaveral pier. The USNS Sentinel had all the equipment on board to work with the submarine, the Frances Scott Key, that was within missile range to Antarctica that was 7,000 miles away.

The timing was great. The day before, the Range Sentinel had an engine breakdown and would not have been able to contribute to the mission. As Captain on the Redstone, I went over to the Navy Commander who was in charge of all the equipment on board. I told him to have his Admiral call my Admiral. Between them we could transfer all the equipment from the Range Sentinel to the Redstone overnight with help from his men, the Navy and shore-side help, using forklifts and front-end loaders. We got the OK and everyone started transferring the equipment, working through the night. It was completed with equipment hooked up and ready to go by the next morning.

There were 250 passengers that were going to board our ship instead of the Range Sentinel. Admiral Bump was in charge of the operation on the submarine side and we had retired Admiral Arnold Schade in charge on our side. Admiral Schade had been the XO of the submarine Wahoo during World War II. During combat Captain Schade was wounded above in the conning tower. He hollered, "Take her down," as he quickly closed the hatch which saved the boat. The XO Arnold Schade became the officer in charge and brought the sub home. I knew Admiral Schade personally because he was in charge of the Cultural Center in Port Charlotte, Florida, where I have lived for more than 50 years.

Our operation went off smoothly and the missile shot went down range to Antarctica. We had 250 guests aboard that day and our Chief Steward had a luncheon of sandwiches with coffee and tea for everyone. This proved if everyone works together, there can be a successful missile launch even with unforeseen complications. Most importantly, the submarine was already out there for the launch. So, another successful mission was accomplished without delay. Admiral Bump was pleased that everything went off without a hitch.

Alter Elementary School

During my 12 years as Captain, I wrote to different schools. One school was the 7th grade class at the Alter Elementary Catholic School in Rossford, Ohio, which is just outside of Toledo. Alter received letters and packages from different parts of the world. My great niece, Jennifer Basinger, attended Alter School and asked me to write to her class. Her teacher's name was Patti Irons. I sent commemorative coins of each space shuttle, plaques from ships, Redstone mugs, and a video tape about Naval training. Many other items were given to the ship by owners of stores, which were sent to the children at the school.

My retirement was scheduled for July of 1987. Before school ended, while visiting my sister, Barbara and her family in Ohio. I decided to visit the school, which was just a short distance away, before I retired. I went to the school and I spent more than an hour and a half at the school. All the students had a great time asking questions about what life was all about at sea for long periods of time. I had a great time through the years talking with the kids.

The Redstone worked with NASA from 1969 to 1987 with the Space Program. I eventually retired from the USNS Redstone on July 7, 1987. I had a glorious career on the ship.

An Interesting Story

The USS America was built and launched in Germany in 1905 for the Hamburg America Lines. The steamer entered service in the Autumn of 1905. She was in the area of the sinking Titanic and had transmitted a message about icebergs where it was sunk 3 ½ hours earlier. After WWI the ship America (German spelling - Amerika) was taken over by the United States. She transported over 51,000 troops back home from Europe and was in the mail service until 1931. She ended up being laid up for 9 years and was reactivated by U.S. Army Transport Service which later became MSTS and then MSC, Military Sealift Command, in 1972. It is interest that I was the son of Seaman John Anthony Arens who was in WWI and ended up in MSTS the same as my father who was on the Edmund Brooke Alexander, formerly named the Amerika. I did not sail on it because at that time I was on the USNS Upsure, another troop ship of MSTS. The Edmund Brooke Alexander was built in 1904 and finally scrapped in 1957.

A Military Mustang

Chapter 30 – MIDDLE EAST

Desert Storm and Desert Shield

I was called back to service in 1991 as the Commanding Officer of the USNS Antares, one of the largest Naval ships during Desert Storm and Desert Shield. It was 941 feet long, 105-foot beam, 120,000 horsepower, 34 knots cruising speed, 27 knots loaded with 59,000 tons. When we arrived off the coast of Daman, we had orders to proceed into the channel to pick up a pilot and then bring the vessel to the dock. Visibility was almost zero because of the 600 burning oil wells that caused fog-like conditions in the whole area. We were moving very slowly at 15 RPM's awaiting the pilot to board our ship. Over the ship-to-shore radio came the message, "Antares, go to anchor! Go to anchor!"

Now we are in a quandary. We couldn't see, and we knew we were too close to other ships at anchor. I took over from Mr. Strasser who was the First Officer and swung the ship to the right out of the channel into the anchorage area. Through the very poor visibility I saw what looked like a structure for an oil well that turned out to be a ship at anchor with his stern facing our ship. I was committed at 15 RPM's. That was the slowest as our ship could go and still be maneuvered. Unlike land vehicles, the slower a boat or ship is going the less control. Speed is needed to overcome the push and pull of the wind and waves.

As we proceeded, at least five men were on the standby to drop the anchor. I was on the bridge right above them and could easily see and talk to them without raising my voice. We were getting small breaks in the mist around us. I could see the two ships and knew we could not anchor that close to each other. I maneuvered right alongside planning to go around his bow by kicking the stern away from him as soon as our ship cleared his bow. It was a tricky maneuver. I had to move blindly by another ship which was anchored somewhere on the other side, which I could not see. When I came alongside, I could look straight down on deck and see men running back and forth and coming up on deck from below. This being an almost 1,000-foot ship the men on the other ship thought for sure that we were going to ram them and cause them to sink. The crew was foreign, so I didn't know if they understood English. But I shouted down to them that we were not going to touch them, but they were in a state of panic by the that time because we were only 15 feet from the side of their ship.

When our bridge moved ahead of their bow, I kicked the engines ahead to turn the stern away from their shipside and cleared them. When we passed them, I could see we could safely drop anchor. The ship was settling in nicely and was floating free from the bottom. I told the pilot station that we were swinging around and settling in at anchor away from other ships. The pilot came out to the ship by the pilot boat and talked to me. After about 20 minutes he explained the tide was going out and that the ship might go aground. The pilot apologized to me

A Military Mustang

about the mix up and about not coming aboard. He also told me I went by the ship really closely. The captain said he never in his life saw ship handling under those conditions and wanted you to know that he understood English. He was telling his crew what I said about not hitting them.

Later while we were anchored, the tide went out and we were sitting on the bottom. That meant I had to send a message informing my boss and everyone else including the Chief of Staff in Washington, DC. I knew that it had to be done or I would be in deep trouble. I sent a long message to all explaining that I was commanded to go to anchor.

I sent another hot message to Washington informing them that I only went aground after the tide went out. When the tide came in, the pilot took our ship to the pier and we had a diving team inspect the bottom of the ship. They found out that there was no damage. Admiral Smith came aboard my ship while at anchor. I met him and we went to my stateroom and I asked him if I could put on my uniform. He said, "Of course." I went and put my dress whites with all my ribbons and medals. When I came out, Admiral Smith and his second in command took one look at me. The whole conversation was about my Ranger Badge and Wings, my Navy Diving Officers Badge, the Navy Parachute Wings, and especially my Combat Infantry Badge. He commended me and gave me a medal for the way we handled the ship under trying conditions.

Saddam Hussein

Saddam Hussein was attacking Kuwait, which is south of Iraq. As you probably know, Kuwait is a small country that was viciously attacked by Saddam's troops. The troops were killing storekeepers by the hundreds and stealing all the gold that was in the stores. They were brutal, using electric drills to torture storekeepers. They would ask storekeepers where they were hiding the rest of the gold. If they didn't tell, the troops used electric drills to make them talk. But the Kuwaiti's were smart. They kept the injured and dead bodies until the Americans got there for proof of how cruel Hussein's men were.

The Iraqi troops loaded vans, cars, and trucks with orders to bring the gold back to Saddam Hussein. There is only one road from Kuwait to Baghdad and they were strung out for a hundred or so miles. Our A10 aircrafts had Gatlin guns in the front and bombs under their wings. They lined up with many other planes and followed them up the road with devastating fire power and sent the Iraqi troops to hell by the hundreds. Not one ounce of gold got to Baghdad. I am sure the flames were so hot that the gold was melted.

High-Speed Boats

During our trip through the Straits of Hormuz we had two very fast high-speed boats headed right for the middle of our ship. They would have caused considerable damage if they had hit us. We had a heavily armed SWAT team aboard for this very thing. We were all set

A Military Mustang

to blow them out of the water when all of a sudden they turned and went under our stern……..they were only carrying goats.

If I had my team open fire on them, it would have been an international incident. The USS Cole had a hole blown in her side because they came alongside, saluted the men on deck, and who then blew themselves up, damaging the USS Cole. Plus, many of the men were killed, so we were not going to take any chances. I had one of the survivors on my ship. He had left the Chief's Quarters and went aft to do some work in his office which saved his life. This incident happened while we were heading back out of the Straits of Hormuz.

Iwo Jima Means Sulphur Island

When I was a parishioner at Faith Lutheran Church in Punta Gorda, Florida, I had many friends in the church. One of my best friends, Seaman James Park, was at Iwo Jima on the Navy ship, USS Bauer. Just before the Marines went ashore the commanding officer took the ship, a destroyer, close ashore and ran as close into the island as possible. They fired all the guns on the shore close as they could before the Marines came in. James told me personally that he witnessed the first flag raising from the ship. We all know the second flag raising photographed by Rosenthal that became famous. I was a speaker at his funeral at our church and I brought this part of his life to the church people.

My other great friend was Tom Whipple. He was one of the Marines that went ashore on that first day,

February 14, 1945, and spent the whole 38-day battle there. Over 6,000 men died and 17,000 were wounded. Somehow, he survived. Over 21,000 Japanese soldiers died for their emperor. It was one of the battles for the island that was used as landing fields for B-29 bombers that were destroying Japanese cities at the end of the war. When I visited Tom Whipple at the end, Sulphur took his life within a matter of two days. He said to me, "sand finally got me." His lungs were bad at that time and he developed pneumonia. He was my hero.

A Military Mustang

Chapter 31 – NOW

Doolittle Raiders

On March 24, 2011, three Doolittle Raiders arrived at the Heritage Museum at Fisherman's Village in Punta Gorda, Florida. Two of the Raiders could not make it because they were not in good health. U.S. Army Col. Richard E. Cole, 95 years old was Doolittle's Co-pilot; Major Tom Griffin, 94 years old was on Plane 9; and Sergeant Dave Thatcher, 89 years old was an engineer gunner. Sixteen U.S. Army Air Force B-25 Mitchell medium bombers were launched from the USS Hornet deep in the western Pacific Ocean. The plan called for them to bomb military targets in Japan. Up to 250,000 Chinese civilians were massacred by the Japanese army in eastern China for helping the Raiders in their retaliatory measures.

I was one of the volunteers in the Museum that day and personally talked to the three men. I helped them get seated at a table in the back of the Museum. They signed autographs for me and many other people that day. I felt honored to have met these heroes who did so much for our country. We really showed Japan that their country could be attacked. Millions of Americans cheered in the streets when President Roosevelt told them about what these men did to Japan. It was the lift our country needed after the devastating attack at Pearl Harbor. I cherish the pictures they signed for me.

General Westmoreland

I was on an airline passenger plane dressed in my high collar whites with my World War II ribbons, Army Ranger ribbons, and Korean War ribbons. After our beloved General Trapnell left us, Colonel Westmoreland moved up to take his place. Westmoreland made Brigadier, One Star General at 37 year of age. We called him the movie star general as he was tall, stately, and had fine-cut features.

The stewardess told me that retired General Westmoreland was on the plane. He was sitting in the coach section, near first class. When the aircraft landed and stopped at our area ramp, I was in the aisle coming forward. The General was standing at his seat bent over a little. As I was moving forward, we all stopped and waited for the ones ahead of us to move. I was looking at the General. He looked at me and said, "What are the CIB, Jump Wings, and Korean ribbons doing on your uniform?" I replied, "Sir, I served under you in Korea and Japan as a sergeant in the G Company under Captain Jonas Epps after coming into your outfit from the 3rd Rangers up on line. Then I went to the Navy Diving School and became a Diving Officer in the Arctic and ended up as a Captain." He looked at me and everyone on the plane, "Can you believe that from an Airborne Sergeant to a Captain!" Everyone started clapping.

Then the line started moving. When we got off the aircraft, he shook my hand said, "I am proud to have a sergeant in my outfit (187th Regimental Combat Team) make captain on a Navy ship." I said to him, "I was proud

A Military Mustang

to serve under you, General." We parted with me saluting him.

My Twilight Years

Now that I am in my twilight years, I enjoy looking back on the years I served my country. I am over 90 years old and now enjoy attending functions for veterans and giving presentations to schools. It gives me great pleasure to dress formally in my clean white Merchant Marine uniform—adorned with the ribbons and medals I've earned from my thirty-seven years of military service. With my lovely wife Lois holding my arm, I think, it doesn't really get any better than this. Without the insistence, persistence, and patience from Lois, this account would certainly never have been possible. My friend and author-navigator Charley Valera has been the missing link to make these pages finally come to life. His tweaking, reviewing, and insight to the book world has been invaluable. For that, I'm forever grateful to the two of them. He helped the pages come alive. All the events and stories here are true according my memory. I wish to apologize, if I mixed up any names or dates after these many years.

I now love going to various military museums. I volunteer at the Heritage Military Museum and I am privileged to meet many other veterans, as well as, people from all over the world – United States, Europe, South America, Australia to mention a few. We show them photos and information about the Civil War, World War I, and other wars up to the present. I also visit grade schools, nursing homes, other war museums, and the Sarasota Military Cemetery. I am writing this book to show what it took to serve my country and to share how it feels to lose best friends. At the same time, understanding my life has always been a bit different than most others. While most families had Christmas dinner together, I was patrolling the seas in search of an enemy of the United States. While most families spent two weeks vacationing together during the summer, I was diving the cold waters of Greenland. You've read the list; it amazes me still. In hindsight, I know it was the life I was meant to lead. My calling, as they say. My hope and blessing for you reading this is that you find your calling and help whenever you can.

Captain John W. Arens
http://navylog.navymemorial.org/arens-john-0

A Military Mustang

NAUTICAL GLOSSARY

Abreast: Situated nearby and to the side of a ship.
Admiral: A commissioned officer in the navy with a rank above vice-admiral, designated by four stars in the US Navy.
Adrift: Without power, oar, or sail to provide control to the ship.
Aft: Near, toward, or at the rear of the ship.
Aircraft Carrier: A navy ship with a landing deck for launching and recovering aircraft.
Anchor Bell: A warning bell mounted on deck forward and rung while at anchor in fog.
Arctic Ocean: The North Polar ocean, mostly covered by a thick ice cap.
Astern: Behind or off to the side near the rear of the ship.
Astronavigation: Celestial navigation or fixing position by sightings of celestial bodies.

Battleship: The largest and most heavily armed classes of warships.
Beam: The width of a ship at her largest part.
Blast: The signal of a ships whistle.
Boatswain: A petty officer in charge of maintenance of a ship's equipment, especially the sails, rigging, and hull. The leading seaman in charge of supervising the crewmen when performing work on deck.
Bow: The forward part of a ship. The foremost outboard part of a ship.
Breach: A hole or opening broken in a hull through which water will enter.
Buoy: A floating aid to navigation, which means a float bearing a number and color for identification which is anchored to the bottom to mark the sides of a navigational channel or location of an obstruction.

Cadet: A student in training at a naval academy.
Channel: The deepest part of a waterway, usually marked for safe passage of water traffic.
Captain: The operator or chief officer in charge of a vessel.

Draft: The vertical distance between the waterline and the deepest part of the keel

First Mate: The deck officer who is second in command after the captain.
Fjord: A body of water which is a narrow inlet bordered by steep highlands and believed to have been carved by ancient glaciers.
Fleet Admiral: The highest-ranking officer in the navy, whose insignia is five stars.
Foghorn: An acoustical signaling device, operated at regular intervals, which provides a loud sound to indicate the position of a ship or lighthouse in the fog or when visibility is obscured.

Ice Breaker: A ship specifically designed with a powerful propulsion and stalwart enforcements to the bow and underside so that it can run through ice fields, breaking up the ice.

Landing Craft (LST): A naval vessel that can run onto the beach to allow soldiers and cargo to disembark.
List: To lean to one side because of improper loading, damage to the hull, or severe weather conditions.

Merchant Marine: The body of commerce that involves the shipping business and commercial ships.
Moor: To secure a boat to a mooring or permanent anchor.
MSC: Military Sealift Command.
MSTS: Military Sea Transport Service.

A Military Mustang

Mustang: A warrant officer or commissioned officer who has been promoted from the enlisted ranks in the US Navy. Generally, officers are college educated and entered service to become an officer.

Petty Officer: A senior enlisted man in the US Navy or Coast Guard in grades E-4 thought E-9.
Port: Referring to the left side of the vessel when seen by someone facing the bow.

Quarter: The stern part of a vessel on either side of the rudder.
Quartermaster: A navy petty officer responsible for the ships steering equipment and the signals.

Rear Admiral: A two-star naval officer.
Rudder: A submerged plane mounted astern for steering a vessel.

Sextant: A precision instrument used to measure the elevation of celestial bodies above the horizon as a means of determining the position of the ship.
Shellback: A veteran sailor, usually referring to one that has crossed the equator.
Ships Bell: The bronze bell that was used to signal the time and change of watch on board a ship.
Stern: The aftermost part of the hull on a ship, referring to the feature between the after end of the keel and the dock.

Warrant Officer: Any officer in the US Navy W-1 though W-5, superior to all enlisted grades.

Hard copy reference for footnotes and bibliography:
Seatalk, The Dictionary of English Nautical Language, www.seatalk.info, keyword (A Military Mustang), published by Mike MacKenzie, Nova Scotia, 2005.
Digital copy link:
http://www.seatalk.info/db/db.cgi?uid=default&Term=(insert the term here)&view_records=yes

A Military Mustang
Ships I Served On

John W. Arens — Military Sealift Command, Atlantic

Ship	Length	Dates	Voyage Begins	Voyage Ends	Rank
ESSO Ships		1944-1950			
USNS R.M. BLATCHFORD	522		1963 Dar-es-Salaam	New York	Jr. Deck Officer
USNS UPSHUR	533	1964-1965	New York	New York	Jr. Deck Officer
USNS CAPISTRANO	205		1965 New York	New York	3rd Officer
USNS REDBUD	180	1966-1968	New York	New York/Newark/Bayonne	2nd Officer/1st Officer
USNS LYNCH	209		1969 New York	Key West	2nd Officer
UNSN GILLIS	314		1969 Key West	Key West	2nd Officer
USNS REDBUD	180	1969-1970	Bayonne	New York	1st Officer
USNS VANGUARD	524		1972 Port Canaveral	Port Canaveral	3rd Officer
USNS WYMAN	286		1972 Charleston	Lisbon, Portugal	1st Officer
USNS LYNCH	209		1973 Lisbon, Portugal	Norfolk	1st Officer
USNS MOSOPELEA	305		1974 Roosevelt Rds	Mayport, Florida	2nd Officer
USNS MIRFAK	265	1974-1975	Savannah, GA	New York/Bayonne	1st Officer/Master
USNS RIGEL	502	1975-1976	Norfolk	Norfolk	1st Officer
USNS MIRFAK	265		1976 Bayonne	Bayonne	Captain
USNS BARTLETT	245	1976-1977	Washington, DC	Charleston	Captain
USNS MIRFAK	265	1977-1978	Sunnypoint	Brooklyn	Captain
USNS WILKES	285		1978 Karachi, Pakistan	Colombo, Sri Lanka	Captain
USNS WYMAN	286		1979 Mayport, Florida	Bermuda	Captain
USNS PVT. JOHN R. TOWLE	455		1979 Balboa, Panama	Port Hueneme, CA	Captain
USNS WYMAN	286		1980 Boston	Cape Canaveral	Captain
USNS KANE	285		1981 Plymouth, England	Bayonne	Captain
USNS RIGEL	502		1981 Subic Bay, P.I.	Norfolk	Captain
USNS HAYES	246		1982 Baltimore	Bayonne	Captain
USNS RIGEL	502		1982 Norfolk	Naples, Italy	Captain
USNS WYMAN	286		1982 Las Palmas, Canary Is.	Tampa, Florida	Captain
USNS HARKNESS	394		1982 Gulfport, Mississippi	Little Creek	Captain
USNS VANGUARD	524	1983-1984	Port Canaveral	Holy Loch, Scotland	Captain
USNS HARKNESS	394	1984-1985	Piraeus, Greece	Mombasa, Djibouti	Captain
USNS LYNCH	209		1985 Cheatham Annex, VA	New London, CT	Captain
USNS REDSTONE	524	1985-1987	Port Canaveral	Halifax/Port Canaveral	Captain
USNS ANTARTES	946		1991 Wilmington, NC	Desert Storm	Captain

Praise for *A Military Mustang*

"Captain John Arens and Charley Valera, two extraordinary men! The first a consummate warrior whose story needs to be told. The second an author so very obviously skilled at gaining and documenting all the phenomenally interesting details. Very well done, Gentlemen!"
    1st Lt. James L. DeVoss USAF (Ret.) F-105 Strike Pilot.
   Distinguished Flying Cross, Three Air Medals, Purple Heart

"*Captain Arens' story speaks of his bravery and dedication to our country. A must read for all.*"
      Richard S. Adams, Retired Police Commander

"As a 20-year Navy veteran, Captain Arens' military accomplishments make him a rare Mustang. I would have been proud to serve with him"
      Douglas Wood; EM1 SS, USN (Ret)
       Submarine USS Tinosa SSN 606

"*An amazing story about an amazing American.*"
    Andy Biggio, US Marine (ret) Author, The Rifle

"*It was fascinating to read the first-person account of Captain Arens that has spanned generations and describes world altering events. Our country is lucky to have people like Captain Arens to protect our freedoms.*"
      Daniel Baus, U.S. Firefighter, Illinois

"*With all his exploits, Captain Arens is the most unostentatious gentleman you could ever hope to meet. He is a true patriot worth reading about.*"
    Deb Allen, Daughters of the American Revolution

"*When I first saw John at Peace River Wildlife Center in his white Navy uniform, he looked so Presidential. He shared such wonderful stories that make you want to hear more. Read his autobiography to hear his stories.*"
            Linda Oneill

Captain John W. Arens
Charley Valera-author of *My Father's War*

John W. Arens and Charley Valera both come from families with military service. Arens, a well-decorated veteran is a career officer, having served over 37 years, including the Army Rangers, Merchant Marines and Navy Seals. Because John never had any formal training to be an officer, he is what's known as a Military Mustang.

Valera's father was a WWII veteran who served in several campaigns across Europe and is the inspiration for his passion for anything related to the military - especially WWII. Charley is a writer, public speaker and general aviation pilot. His first book, "My Father's War: Memories of our Honored WWII Soldiers," has won several literary awards.

Both John and Charley currently live in Southwest Florida.

Charley Valera
Best-Selling Author of *My Father's War*

Author Charley Valera's own father spent almost four years fighting during WWII and lived out the rest of his life "without a story to tell" to his four sons. To learn more about what it was like for his father and those that served, Valera conducted heartfelt interviews with WWII veterans from both theaters of war and all branches of service. He brings the reader into the battlefield, aircraft, destroyer and marching through the country sides. These are the stories that hadn't been discussed in decades. Within these pages are their personal photos and memories of which many of us can call My Father's War.

Future Books by Charley Valera

Veterans Tell the Funniest Stories
Look for Charley Valera's next book that looks on the lighter side of life for U.S. Veterans. While we all know that our veterans have sacrificed greatly to serve and protect our country, there were fun times. Those who have served know, the military is not without laughter. Charley shares some of the most hilarious stories from our veterans. Stories from the front and stories from the home front. "You just can't imagine some of the funny stories the WWII veterans shared with me. But you will," Charley told me when asked to describe his next book.

Working with the World's Top Rock Bands
Join Charley as he tells what it is like working with top rock bands. Learn about the tough times and the funny times. Learn about what went on behind the scenes. Who really made the decisions? Who kept the band together through the tough times? Connect with Charley as he shares the laughter and the work – the inside scoop that you always wondered about.

www.ingramcontent.com/pod-product-compliance
Lightning Source LLC
Chambersburg PA
CBHW062114280426
43661CB00111B/1424/J